DRONES

DRONES

An Illustrated Guide to the Unmanned Aircraft That Are Filling Our Skies

Martin J. Dougherty

amber
BOOKS

Editorial and design by
Amber Books Ltd
74–77 White Lion Street
London
N1 9PF
United Kingdom
www.amberbooks.co.uk
Appstore: itunes.com/apps/amberbooksltd
Facebook: www.facebook.com/amberbooks
Twitter: @amberbooks

ISBN: 978-1-7-8274-255-5

Project Editor: Sarah Uttridge
Picture Research: Terry Forshaw
Design: Hawes Design

Printed in China

Contents

Introduction

Not very many years ago, few people had even heard of drones. Most of those who had would probably have an idea from science fiction or techno-thrillers about what a drone was and what it might be capable of, but no real knowledge. Yet, in just a few years, drones have gone from obscurity to near-constant media attention. We hear of drone strikes and drone surveillance in the world's trouble zones, and of drones delivering packages – even pizza – in the commercial world.

A surprising number and range of users have been operating drones for some time, although the rest of the world has known little about it. Outside of the military, drones have been used for research purposes, or to monitor the environment. Commercially available drones can now be bought at quite a cheap price by private users for recreational purposes.

Yet, in truth, there is nothing really new about the idea of a remotely operated vehicle. The word 'drone' has entered the popular vocabulary, but long before this happened users were flying remote-controlled aircraft and helicopters, or racing radio-controlled cars. Remotely controlled weapons have been in use for several years – although not always with a great deal of success. It is, however, debatable whether these were, strictly speaking, drones.

What is a Drone?

One useful definition of a drone is a pilotless aircraft that can operate autonomously, i.e. one that does not require constant user control. This means that traditional radio-controlled aircraft

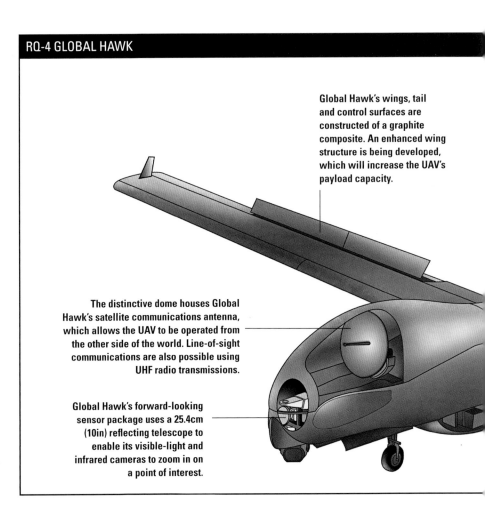

RQ-4 GLOBAL HAWK

Global Hawk's wings, tail and control surfaces are constructed of a graphite composite. An enhanced wing structure is being developed, which will increase the UAV's payload capacity.

The distinctive dome houses Global Hawk's satellite communications antenna, which allows the UAV to be operated from the other side of the world. Line-of-sight communications are also possible using UHF radio transmissions.

Global Hawk's forward-looking sensor package uses a 25.4cm (10in) reflecting telescope to enable its visible-light and infrared cameras to zoom in on a point of interest.

and the like are not, in the strictest sense, drones. Nor are many underwater Remotely Operated Vehicles (ROVs), and not only because they are not aircraft. In fact, many recreational 'drones' are not really drones, as they are semi-autonomous. However, it is useful to widen the definition of a drone somewhat in order to cover a range of similar vehicles that undertake the same role using broadly the same principles.

Right: Operating a UAV is a complex business, which has been described as similar to flying a plane whilst looking through a straw. In addition to piloting the vehicle, operators must control cameras, radar and other instruments, and hand-off data to other users, making the operation of a large military UAV a multi-person task.

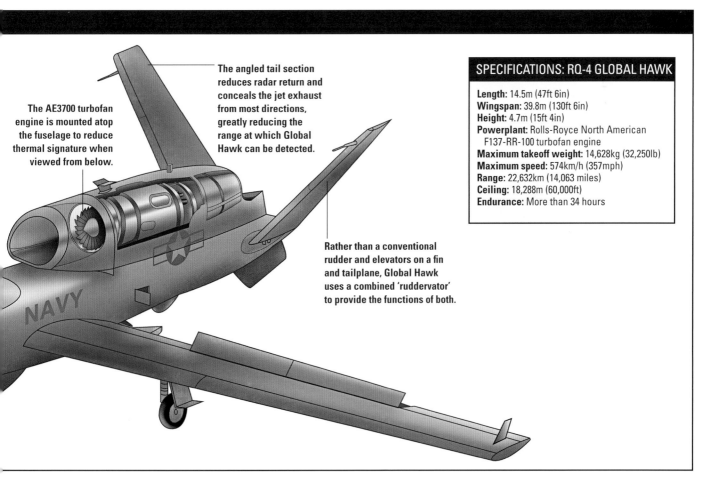

The AE3700 turbofan engine is mounted atop the fuselage to reduce thermal signature when viewed from below.

The angled tail section reduces radar return and conceals the jet exhaust from most directions, greatly reducing the range at which Global Hawk can be detected.

Rather than a conventional rudder and elevators on a fin and tailplane, Global Hawk uses a combined 'ruddervator' to provide the functions of both.

SPECIFICATIONS: RQ-4 GLOBAL HAWK

Length: 14.5m (47ft 6in)
Wingspan: 39.8m (130ft 6in)
Height: 4.7m (15ft 4in)
Powerplant: Rolls-Royce North American F137-RR-100 turbofan engine
Maximum takeoff weight: 14,628kg (32,250lb)
Maximum speed: 574km/h (357mph)
Range: 22,632km (14,063 miles)
Ceiling: 18,288m (60,000ft)
Endurance: More than 34 hours

It is quite difficult to pin down a working definition of 'drone' that does not immediately founder on the rocks of the first exception it encounters. In theory, any remote-controlled aircraft can operate like a drone, inasmuch as it can be pointed in the right direction and set to fly straight and level. During this period, the operator can let go of the controls and the aircraft will go on its way without control input.

This is not really a drone operation, however. To be such, the aircraft would need to be able to make some decisions for itself. A simple autopilot that used the aircraft's control surfaces to keep it on course might not qualify, but one that could be given a destination and fly the aircraft to it, possibly making course changes as necessary, would fit the common definition of a drone.

Some 'drones', especially those operated by the military, are primarily operated by a pilot from a ground station. They can undertake autonomous flight, but are normally under constant control.

Right: The majority of small UAVs resemble the miniature aeroplanes flown by radio-control enthusiasts for many years. There are more design options as size increases. The RQ-2 Pioneer at centre rear is essentially a conventional small aircraft whilst the RQ-15 Neptune at right rear is a flying boat designed to land on water.

Military drones, such as Predator, require considerable piloting skills and push the boundaries of what is, and what is not, a drone. Indeed, many operators dislike the

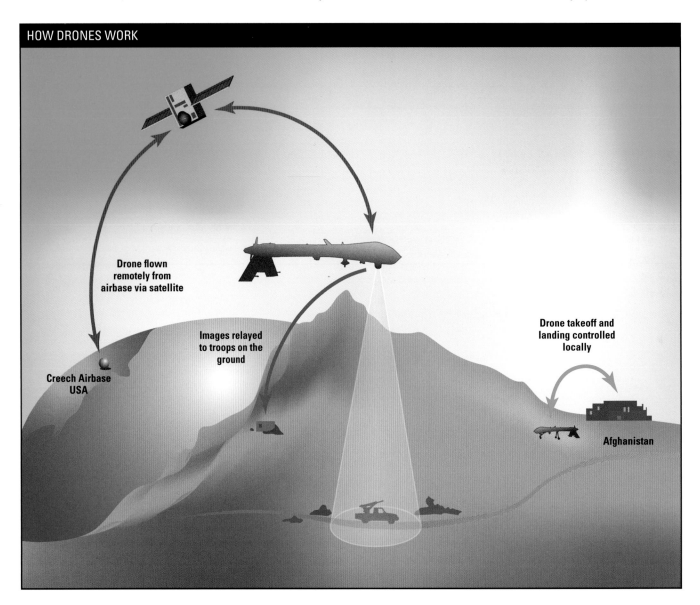

HOW DRONES WORK

Drone flown remotely from airbase via satellite

Images relayed to troops on the ground

Drone takeoff and landing controlled locally

Creech Airbase USA

Afghanistan

AUTONOMOUS LANDING PROCEDURE

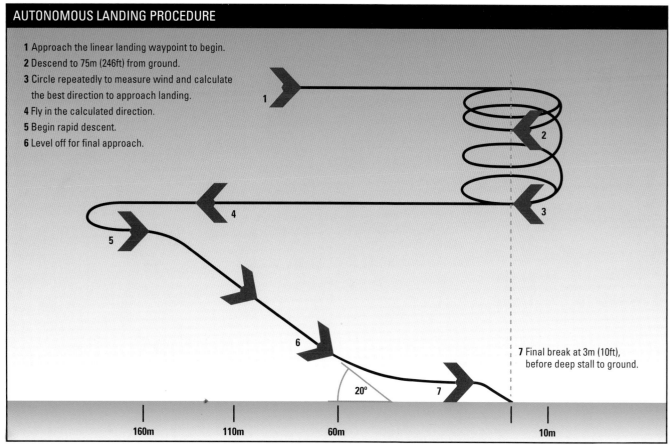

1 Approach the linear landing waypoint to begin.

2 Descend to 75m (246ft) from ground.

3 Circle repeatedly to measure wind and calculate the best direction to approach landing.

4 Fly in the calculated direction.

5 Begin rapid descent.

6 Level off for final approach.

7 Final break at 3m (10ft), before deep stall to ground.

20°

160m   110m   60m   10m

use of the term 'drone', as what they do is every bit as difficult as flying an aircraft that they are aboard.

A US military Predator drone, flying under constant control by a ground operator, is not really a drone under the strict definition above. It is an Unmanned Air Vehicle (UAV), a term that its operators prefer in any case. Similarly, a remotely operated vehicle used for underwater operations – say inspecting deep-water pipelines – may be under constant control and would best be considered a Remotely Operated Vehicle (ROV) or an Unmanned Underwater Vehicle (UUV).

Missiles and torpedoes meet the definition of a drone in many ways. They can guide themselves, making decisions about flight path or direction of travel, and often operate autonomously. Some are manually guided or home in on a manually controlled targeting system, such as a laser designator. Missiles and torpedoes are not, however, normally considered to be drones, even though there are drones that fulfil a similar role.

Likewise, it can be difficult to decide whether a given vehicle is a drone or not by just seeing it in operation. A drone-like aircraft could be under constant manual control, perhaps by using a First-Person View (FPV) system. This is essentially a camera in the front of the vehicle that gives the operator a pilot's view. A small aircraft-type 'drone' might be a traditional model aircraft, or might be flying under its own control using GPS guidance.

Thus the field of drone operations is rather complex and overlaps in some other areas. For our purposes it must suffice to use a fairly loose definition of the term 'drone'. We shall therefore consider a drone to be any vehicle that has no pilot aboard, and which is capable of at least some autonomous functions that require onboard decision-making, and which does not obviously fall into some other category, such as a missile or a guided artillery shell.

Historical Attempts

Historically, there have been numerous attempts to create pilotless vehicles,

mainly for military purposes. Among the more ludicrous was the idea of using 'organic control' in a missile. The organic control took the form of a pigeon trained to recognize a particular type of target and peck at it. The intrepid pigeon could then guide a missile to the target by pecking at a screen in the front of the missile, which was connected to the controls to allow corrections. With the missile centred on the target, the pigeon's pecks would keep the missile on course; deviation would be corrected as the pecks moved further from the centre of the screen.

The development of electronic systems small enough to fit inside a missile ensured that organic control was abandoned in the early 1950s. It has not, strangely enough, been revisited.

Other attempts at creating autonomous vehicles, also originating in World War II, were more straightforward. The V1 'flying bomb' was essentially a pilotless aircraft powered by a pulse-jet engine. It was of simple construction and cheap to build, but carried a significant warhead. The

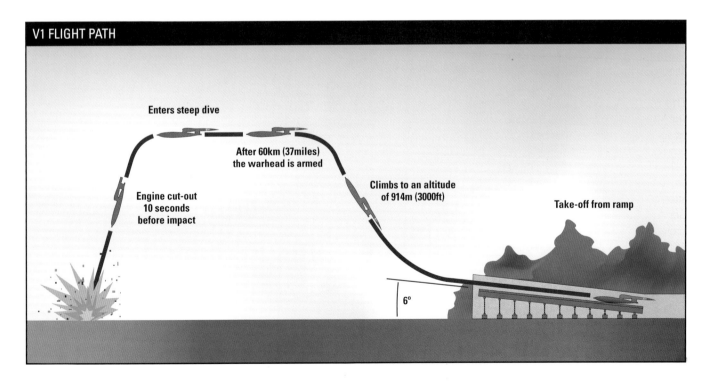

V1 FLIGHT PATH

Enters steep dive

After 60km (37miles) the warhead is armed

Climbs to an altitude of 914m (3000ft)

Engine cut-out 10 seconds before impact

Take-off from ramp

6°

V1 had an autopilot of sorts, which kept it level, and a very simple inertial system that activated the dive mechanism. A small propeller on the front of the V1 was driven round by air passing over the nose. When the propeller reached a preset number of revolutions, the bomb would, in theory, have travelled the requisite distance and should begin its dive towards the target.

In practice, the careful calculations used to determine the dive point were thrown out by headwinds, tailwinds or imprecision in the device, causing many V1s to fall long or short. A side wind would also send the device off course. It had no navigation system as such, merely being launched in the direction of the target. This also made the V1 vulnerable to interception by aircraft and ground-based guns, as it had to follow a straight course.

Lack of a true autopilot allowed the flying bomb to be defeated by either flying ahead of it so that the slipstream of an aircraft caused it to veer off course and crash, or by the rather more hands-on approach of tipping the wing while flying alongside. The warhead did still come down somewhere, but it could be sent off target and prevented from striking a densely populated area.

The Mistel

The V1 flying bomb was not a drone as such, nor was it a missile. It can be considered a precursor to the modern military drone and missile, and helped prove the concept was workable. Another attempt at the same goal was the German 'Beethoven Device', a composite aircraft created from a single-seat fighter riding atop a larger aircraft. The latter, known as the 'Mistel', was packed with explosives and flown by remote control from the smaller aircraft.

The Mistel was usually a light bomber, such as a Ju-88, although various aircraft were used. Often these were obsolete designs, the intent being to obtain some

Above: The V1 was a rather primitive forerunner of the modern cruise missile. Its effectiveness was limited by the lack of a remote piloting capability or an effective automatic navigation system. The technology to implement these systems had not yet been invented, which was fortunate for the intended victims.

useful service from aircraft that otherwise would have to be retired. In theory, the composite aircraft was flown close to the target, after which the fighter would escape while the Mistel component delivered its large warhead to the target. In practice, although over 200 of these devices were built, few achieved any useful results.

The final development of the Mistel concept was to have used jet aircraft that were emerging at the end of the war. This was a rather wasteful deployment of advanced combat aircraft, especially since the Mistel could deliver a relatively limited warload, and

could, of course, do so only once. The composite aircraft was also very vulnerable to interception.

Although not a drone in the true sense, the Beethoven Device was a credible attempt – given the technology of the day – to create something in between a guided missile and a one-use attack drone aircraft. It was ultimately a failure, but demonstrated the concept of a pilotless aircraft capable of – in theory at least – undertaking a combat mission.

Other wartime projects included radio-controlled glide bombs and acoustic homing torpedoes, which had at least some of the characteristics of a drone weapon. Although these were of limited impact during World War II, they did work, and well enough that the technologies were further explored. This led to the development of modern torpedoes and guided missiles of various sorts, whose technology would pave the way for drones. Without the work done on missile guidance just after World War II, today's drones would not be possible.

The idea of a remotely piloted vehicle continued to be developed after the war, not always by the military. Radio guidance systems used in glide bombs and primitive missiles were eventually developed to the point where civilian enthusiasts could fly model aircraft and helicopters for recreation. The affordable

Above: The Mistel concept was an attempt to create a guided weapon using manual control. The pilot detached his small aircraft from the larger, explosives-packed carrier and guided its final approach to the target by remote control. Attacks were made by this method, though with limited success.

Left: By the mid-1950s, radio control of model aircraft was common enough that contests like this one could be held. The development of remote control filled in one of the missing elements that led from crude pilotless flying bombs to modern UAVs.

radio-control hobby expanded to include boats and racing cars, and today it is sufficiently cheap that radio-controlled children's toys are commonplace.

None of these developments created a drone, as such, but they were all steps towards that goal. Ever more powerful miniaturized electronics made it possible for a model aircraft to carry a fairly powerful processor – one that could handle decision-making if need be. This made it possible for a true drone to be created; one that could be instructed with what to do – for example, the start and end points of its flight path and a few waypoints in between – and left to make its own decisions about exactly how to carry out these instructions.

Advances in materials technology were important to the creation of workable drones, as well as to the aircraft industry in general. Lightness and strength are crucial when building any air vehicle; weight saved from structure can be used for payload. This is particularly important with small drones, where just a few grams can make the difference between being able to carry a useful payload and not getting off the ground at all.

A greater choice of materials is possible with smaller drones that do not have to carry humans, cargo and the heavy engines required to provide enough thrust to get off the ground. It would simply not be possible to build a full-sized aircraft out of some of the materials used to produce drones. The stresses experienced by a small drone during flight or landing are totally different from those encountered by larger aircraft, but as size increases, a drone must be constructed more like a conventional aircraft.

The introduction of the Global Positioning System (GPS) was another

Below: Missile guidance drove the development of many systems now used in UAVs. The AIM-9 Sidewinder missile first saw action in 1958, but did not always meet expectations. This test in 1974 at Point Mugu was part of ongoing development that resulted in the AIM-9H model, the first with solid state electronics.

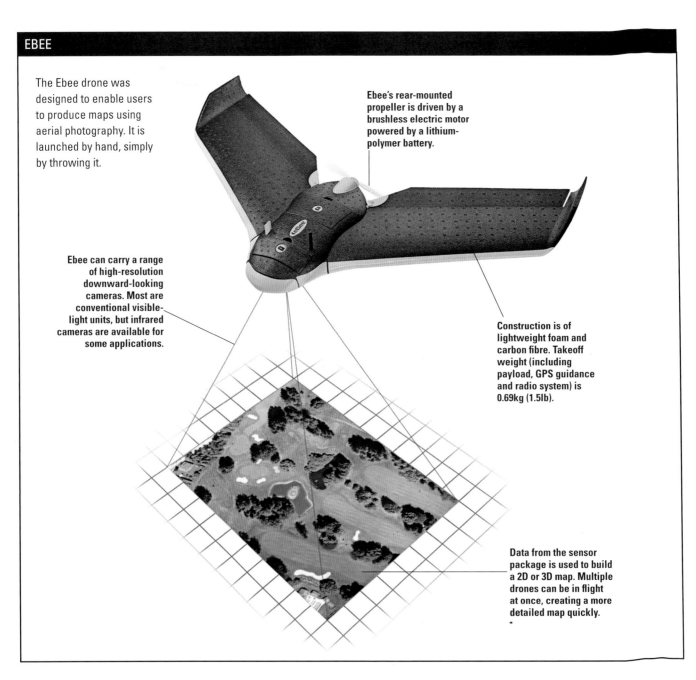

The Ebee drone was designed to enable users to produce maps using aerial photography. It is launched by hand, simply by throwing it.

Ebee's rear-mounted propeller is driven by a brushless electric motor powered by a lithium-polymer battery.

Ebee can carry a range of high-resolution downward-looking cameras. Most are conventional visible-light units, but infrared cameras are available for some applications.

Construction is of lightweight foam and carbon fibre. Takeoff weight (including payload, GPS guidance and radio system) is 0.69kg (1.5lb).

Data from the sensor package is used to build a 2D or 3D map. Multiple drones can be in flight at once, creating a more detailed map quickly.

critical technology. GPS signals are available to any device that is capable of receiving them, making it unnecessary to carry more than a receiver. This made accurate navigation possible without needing to carry numerous instruments or requiring operator updates, and this was a key factor in enabling modern drones to operate autonomously.

Communications technology also had to advance to the point where systems developed for other purposes could be co-opted for drone operations. Developing a control system purely for the purpose of recreational drone flying would make it a hobby for the very rich only, but today's communications equipment has gone beyond single-

purpose items. Indeed, interoperability and compatibility are strong selling points for tablets, laptop computers, mobile phones and the like.

The availability of general-purpose, programmable communications and computer equipment meant that drone manufacturers could make use of existing technology. For those on a military budget, this was less critical than in the recreational sector, but, all the same, use of Commercial Off-The-Shelf (COTS) components drove down both development costs and the price tag of the final unit. For the recreational and private civilian user, this was the development that made drone operations possible at all.

Thus it can be seen that today's drones – military, commercial, recreational and all other types – were not created out of nothing. They are the end product of a process that began with crude experiments and expedients using whatever technology was available, and later developed into sophisticated systems once the requisite technology became available. Much of this was not specifically created for drone operations, but it was adopted or adapted from other applications.

Now that drone technology has become a mature field, it is likely that development of future systems will be – in some cases at least – specific to the field. This was not possible until it had been proven that there was money to be made from developing new and better drone technology; this is definitely now the case.

Drones are becoming increasingly common in a variety of applications, which is raising some new questions. Is it ethical to mount weapons on an unmanned aircraft? Or to permit anyone to fly a camera-carrying drone more or less wherever they please? What are the implications for national security? For privacy? Are drone-captured images acceptable as evidence in court? What laws need to be enacted to regulate the operation of potentially dangerous flying machines in urban areas?

As with any new technology, society needs time to adjust. Law and social custom must adapt to what is being done rather than attempting to guess what might happen based upon what is possible or foreseeable at the time. It is

not always possible to predict the uses to which an emerging technology might be put, and of course, not all technologies live up to expectations.

What is clear at the present time is that drones have opened up new possibilities in some cases, and in others have made capabilities accessible to private users on a tight budget. It is now possible to spray crops more cheaply with a drone than by hiring a pilot and a crop-duster aircraft. A farmer or conservationist can fly a model aircraft over an area and receive up-to-the-minute images of vegetation or wildlife. Law enforcement agencies can put a camera in the air for a fraction of the cost of a helicopter, and the military can undertake long-range surveillance of an area without the cost and potential consequences of maintaining a manned aircraft presence.

All of these applications – and many others – rely on a range of technologies. Some have emerged in recent years, while others are developments that have been ongoing for decades. The trend towards cheaper, lighter systems continues, which in turn will make more capable drones available at an ever more affordable price.

Drone Technology

The majority of drones are aircraft, or aircraft-type devices. These fall into two broad categories – rotorcraft and winged aircraft. The latter do not necessarily need to be powered – a suitably constructed glider drone could stay aloft for a very considerable time – but most use an engine.

The vast majority of winged drones – and all of them outside of the military – use a propeller to provide motive power. This could be driven by an internal combustion engine, but that is only practical on larger drones. Such drones are also prone to be noisy and may pose a hazard in the event of a crash. Legislation regulating electrically powered drones is far more likely to be agreed

than a body of law allowing private citizens to send airborne fire hazards anywhere they want to – or anywhere their incompetence takes them!

Thus the majority of winged drones use one or more propellers driven by electric motors. The propellers may be mounted in a 'tractor' configuration, i.e. positioned to pull the drone through the air, but 'pusher' types are very common. A pusher propeller is, naturally, located on the rear of the drone or its wings. This has the advantage of leaving the nose clear for either cameras or an equipment bay.

As with all aircraft, landing is a hazardous time for a drone. Propellers can be damaged – and can cause damage – on a poorly executed landing or other collision. However, there are no real alternatives. As already noted, flammable liquid fuel is also a hazard. This is acceptable where the drone has advanced systems to prevent or mitigate a crash and the operator is likely to be extensively trained, such as a military drone system, but for the average civilian user an electric drive is much safer.

Batteries are a relatively heavy component of a drone, but power output and capacity are both improving constantly. The ability to recharge batteries or swap a charged set into a drone for further operations is a significant advantage, especially since many drones are not limited by crew endurance. Where a manually piloted aircraft must return to base due to crew fatigue, a drone can operate more or less endlessly, either autonomously or under control of a series of ground operators.

Similarly, whereas manned aircraft require extensive safety checks and sometimes significant maintenance before each flight, it is usually possible to land a drone, swap out the battery pack and send it off again.

The performance of a drone (or any aircraft) in terms of height it can reach, conditions it can tolerate and speed it

can fly at is defined by the output of its power source. As a rule, delivering a lot of power drains a battery quickly, whereas a lower level of output can be sustained for a much longer period. Power-to-weight ratio is important for high-speed, high-performance drones, but for the most part what matters is endurance.

Drones tend to be light; even quite large military drones are smaller and lighter than a manned aircraft. Control

Left: The Spanish-designed Atlante UAV was created to be capable of operating in civilian airspace. Many drones are prohibited from doing so as they lack automated collision-avoidance systems or fail to meet other aerospace legislation requirements. Atlante thus has a wider range of potential applications than purely military UAVs.

electronics take up far less space than a pilot and other crew, plus all of the supporting systems that they need. A pilot needs space to move around, at least enough to operate controls and get in and out of the aircraft. He needs a seat, a pressurized cockpit and sufficient space between himself and the instrument panel so that he can focus on it. A drone needs none of these things, saving both space and weight.

Small size and light weight have a number of advantages, not least the ability to fly for a long time on little power. However, a lightweight aircraft is more susceptible to wind and the effects of air temperature, humidity and so forth than a heavier one. There is also a tradeoff to be made between payload and capability. Many drones are so small and light that

Above: The Yarara UAV was developed in Argentina for the domestic and export market. Whilst modest compared to cutting-edge UAVs fielded by the USA, Yarara offers the most important features of a drone system – low-cost airborne reconnaissance and surveillance operations. It can be operated from the most primitive airstrip.

fitting a different camera, thus changing weight by perhaps tens of grams, can seriously alter the drone's performance and shorten its endurance considerably.

Winged drones fly like conventional aircraft. Airflow over the wings provides lift, assuming the drone is travelling fast enough. A lightweight drone does not need all that much lift, which is usually an advantage. However, a sudden tailwind can be a real problem for a very light

Above: Rotorcraft drones, such as Octane, offer a highly stable and precisely manoeuvrable sensor platform, at the cost of a high power consumption to stay in the air. They are most suitable for short-range, low-altitude operations and can fly in very cluttered environments where conventional winged drones simply could not.

aircraft, as it can rob the drone of lift and cause a crash.

Stability is usually provided by an aircraft-type tailplane and vertical stabilizer (fin), although drones can make use of the same modern design concepts as aircraft. Thus, in place of a standard rudder on the fin and elevators on the wings and tail, a

drone can use an angled tail section to carry out the functions of both tailplane and fin. Control systems are needed for all moving surfaces, which adds weight and complexity, although modern servos are small and require little power.

The primary advantage of a winged drone is that it requires very little power to keep it aloft. Providing airspeed is above the drone's stall threshold, aerodynamic lift carries it along. This requires far less power than a rotorcraft drone, making for greater endurance or perhaps higher speed for the same power output.

Some rotorcraft drones are built like miniature helicopters, with a single main rotor providing lift. Some means is required to counteract the torque created by the rotors, otherwise the drone would rotate uncontrollably. A tail rotor is the usual solution on a helicopter-type drone, and this provides very precise directional control. However, most rotorcraft drones use multiple rotors.

Multi-Rotor Design

A multi-rotor design (e.g. a quadcopter, with four rotors) has stability advantages over a standard helicopter design, although it does have to carry a motor for each rotor and control systems to operate it. Rotorcraft use a brute-force method of flying, essentially dragging

themselves into the sky using the thrust from their propellers. This means that the entire weight of the drone must be borne by its rotors, whose output must collectively exceed the weight of the drone and its payload.

This high output does drain batteries far more quickly than a winged design and results in a lower speed for the same amount of power, but this is compensated by great precision. A rotorcraft drone can hover; a winged design cannot. This makes rotorcraft drones an excellent choice for many applications, including stable camera platforms and any operation in a tight space. Rotorcraft can be used indoors, for example, whereas a winged drone would be inappropriate.

A propeller-driven drone, whether

rotorcraft or winged, needs a motor to provide power to the rotors. That motor can be electrically driven or use internal combustion. A jet engine is only an option on larger drones, but can deliver sufficient thrust to carry a large payload and/or reach high speeds. This is currently only an option on high-performance military drones, but as the technology matures we may see large cargo-carrying drones enter the commercial marketplace.

There are plans for a pilotless airliner, although it remains questionable whether passengers will accept an aircraft without a pilot. Already, commercial flights make great use of automation, with routine flight operations handled by the aircraft's electronics. However, the jump from automation-assisted piloting to true drone operations is a very significant one.

Below: Like any new technology, UAVs must be integrated into operations with other systems if they are to achieve their potential. This Hunter Joint Tactical Unmanned Air Vehicle operator is taking part in a combat search and rescue exercise. Deploying drones to search for downed aircrew increases coverage without exposing manned aircraft to excessive risk.

Sensor and Communications Technology

Most drones carry at least one form of sensor, i.e. a means of obtaining information, and have a communications system to allow them to receive commands or navigational data. Communications are by radio transmissions, and may be one-way or two-way depending on the drone design.

If a drone is to be programmed to carry out a task (e.g. photographing an area), and can then perform this without additional input, there is no need for two-way communications. The same applies if the drone is to be controlled from the ground by an operator who can see it. A simple receiver is all that is required in this case, although if the drone is to collect data then it must be retrieved. This will usually mean physically retrieving the drone and downloading from its internal storage.

Obtaining data from a remote collector is a rather simpler process than it might have been a few years ago. In the early days of satellite reconnaissance, photographs were taken on conventional film that was then ejected from the satellite and fell to earth. The film canister could be caught in mid-air by a specially equipped aircraft as it descended on a parachute. The advent of digital photography and easy data transmission made this sort of complex and expensive operation obsolete, although many drones use onboard data storage without transmission equipment that must be retrieved for download.

GPS

A drone may make use of the Global Positioning System to navigate. GPS satellites transmit a signal that allows any suitably equipped receiver to work out exactly where it is with a high degree of accuracy. There are limitations to GPS guidance, however. The signal can be lost, and, in any case, the drone does not know what is around it from the GPS

Above: First-Person-View (FPV) drone systems send a live feed from a camera in the nose of the UAV to a display used by the operator. This can be a laptop or tablet screen, but some systems use a set of goggles to give a more immersive and intuitive pilot experience.

signal unless it has been programmed with a map of the area. This has to be a 3D representation, incorporating height as well as location relative to the ground, to avoid accidents and it still cannot prevent collisions with a moving object, such as an aircraft or a bird.

Despite these limitations, GPS guidance is cheap and effective, and may be all that is needed for a simple drone, such as one programmed to photograph an area from the air for the purposes of map-making or environmental monitoring. The user can programme a series of waypoints and send the drone on its way, after which its internal electronics can use the constant updates from GPS to compensate for the effects of wind and other environmental factors.

If GPS guidance is not used, or if direct control is required over greater than visual distances, the drone will need to transmit as well as receive data. A First-Person-View (FPV) drone must transmit images from its camera back to the user, who then sends control signals back. Although lacking some of the stimuli associated with flying an aircraft, such as inertial effects on the pilot and his own sense of balance, FPV can provide an experience much like real flight and is an effective way of controlling a drone once the pilot gains a reasonable level of skill.

Even for civilian operators, there is always a danger of control signals becoming disrupted or interfering with one another. Radio control enthusiasts have for many years used a system of different control frequencies to avoid interference, often denoted by a small flag on their control device antenna. Modern control systems are more sophisticated and do not rely on simply receiving a signal that directly actuates control surfaces.

Today's data transmission equipment is sufficiently sophisticated that control

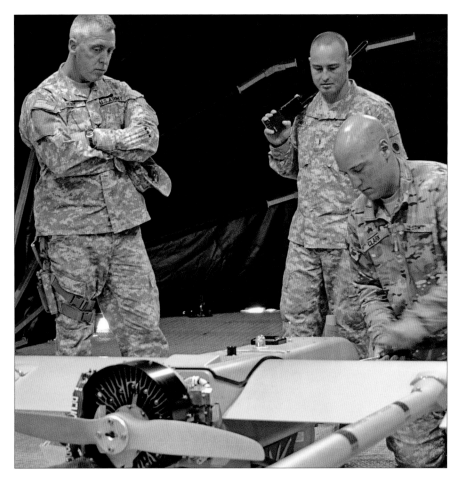

Above: The RQ-7 Shadow UAV is the airborne component of a system that requires vehicles for transport and launch, control stations and, of course, a crew to maintain and prepare the drone for flight. A standard deployment of three ready UAVs and one carried in a disassembled state as a spare requires 22 personnel.

signals intended for some other device or drone will be ignored, but this does not rule out the possibility of accidental interference swamping the signal. For military users, a higher degree of robustness is, of course, necessary. A drone that could easily be jammed by any strong transmission in the area would be of little use.

Fiction is filled with hackers who can achieve incredible things, such as making cars explode by remote control, and the idea of military drones being hacked to place them under the control of the enemy – or someone with an

agenda – is a sufficiently common theme that many people believe it is possible in reality. Military systems are extremely resistant to external hacking; it is unlikely that a 12-year-old with a tablet would be able to gain control of a missile-armed military drone under any given set of circumstances.

This is not least because the threat is recognized and taken seriously by military drone operators. A device that can be controlled remotely is – in theory at least – vulnerable to hijacking by someone else who figures out how to send control signals. Even fairly basic

civilian drones are resistant to such attempts; the budgets available to military drone developers ensure that software is resilient and that the drone cannot easily be retasked to some nefarious purpose.

Most civilian and commercial manufacturers advertise their drones'

capabilities which is generally good for sales. However, the manufacturers of military and some specialized drone systems must tread a fine line between showing what their systems can do and denying competitors or potential enemies useful information that could be used to create countermeasures.

While the capabilities of weapons are often easy to deduce, even if the manufacturer does not advertise them, electronic equipment is often more of a mystery. It can be hard to determine what is under a cowling or in a pod, and, even if a system can be identified, its precise capabilities may not be obvious.

Above: The MQ-1 Predator UAV has a forward-looking camera for flight operations, but its mission-payload optics are housed in a turret that can be controlled independently of the aircraft. For all the media publicity surrounding 'drone strikes' the Predator's most important system is its sensor package.

Other than electronics necessary for communications with the operator and control of the vehicle, drones tend to carry two kinds of electronic equipment. Electronic warfare systems are used only in military and some security drones. They are designed to make use of the electromagnetic spectrum to gain an advantage, or to deny the enemy one, whereas reconnaissance systems are used to obtain information and may be useful in a wider range of applications.

Sensor Systems

The commonest sensors used aboard drones are cameras of various types. Today's digital cameras are small enough to fit in even an extremely small space and still capture high-quality images. Just as importantly, they can store large numbers of images, making possible applications that would not have been feasible just a few years ago. Nevertheless, various types of camera are necessary for different applications.

A constant video stream is necessary for first-person operation of a drone, or real-time observation of a target. This is especially important in military applications, where decisions must be made based on what the target is doing right at this moment. A party of innocent-looking people may suddenly produce weapons, or innocents may wander into a target area just as a strike is about to be launched. Real-time video allows 'man-in-the-loop' decisions to be made, such as pulling a missile off target, launching a pre-emptive attack, or aborting an airstrike that is about to be made.

In non-military applications, real-time video can be used for surveillance, or even film-making, but is often less critical than when deadly force is being used. Real-time observation of a disaster, such as a multiple vehicle crash or a fire or earthquake, can be extremely useful in time-critical operations, such as locating and retrieving casualties or pulling responders back from an imminent threat.

Video is not necessary to many applications, however, and, as a rule, single images can be of higher resolution than a video feed. Thus many drones carry still-image cameras as well as video units. These still-image cameras may be of a specialist type, such as extremely high-resolution or long-distance cameras, or units that are designed to work in extremely poor light.

Thermal imaging and infrared cameras are useful in many applications, but are accompanied by more standard visual cameras. Infrared cameras detect thermal emissions (heat) rather than normal light, and can thus distinguish objects that would otherwise be hard to spot. This is, of course, not infallible; a vehicle that has cooled to the temperature of its surroundings will not stand out on infrared – indeed, it may be all but invisible, even though it is quite obvious to the naked eye.

Infrared can be used for navigational purposes, enabling the drone to see

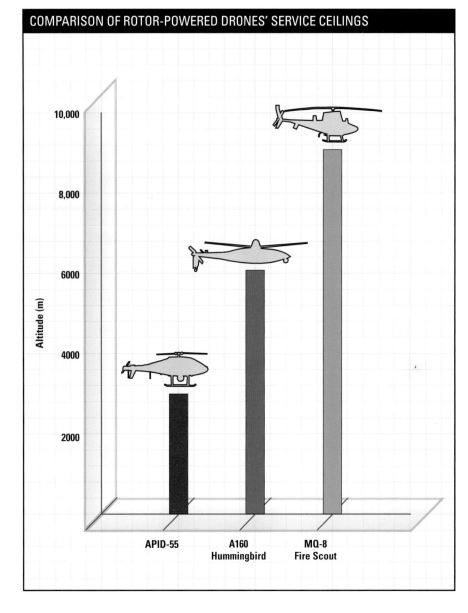

COMPARISON OF ROTOR-POWERED DRONES' SERVICE CEILINGS

Altitude (m)

10,000

8,000

6000

4000

2000

APID-55

A160
Hummingbird

MQ-8
Fire Scout

through haze or rain that might obscure a visual camera, and also has applications ranging from military reconnaissance to examining pipes for flaws. Some infrared cameras (and other types too) are mounted in a sideways-looking position, essentially pointing out sideways from the drone. The camera's field of view sweeps along as the drone moves, and can be aimed by circling around a point of interest.

Alternatively, the thermal camera can be mounted to point straight forward, creating what is known as Forward-Looking Infrared or FLIR. FLIR is highly useful for avoidance of obstructions, but requires that the drone be aimed right at whatever it is observing. A sideways-looking camera mount can be kept on target for an extended period by circling; by definition, a FLIR unit can only be pointed at the target for so long

before the drone has to come around for another pass.

Other thermal cameras are mounted in gimbal or turret mounts, enabling them to traverse and point wherever the operator wants, regardless of the drone's movements. This has advantages, but of course, the mounting adds weight and must be powered, so it is not very useful for small drones. A gimbal or turret mount can carry more than one camera and

PORTABLE CONTROL STATION

The Latvian Penguin UAV can fly autonomously or be controlled from a portable ground station.

Military drone operations take place in a rugged environment where fragile equipment can be damaged. The carrying case provides protection and keeps necessary components from going astray. In addition to the two laptops, the case contains a joystick, mouse, battery packs and antennae for the control system.

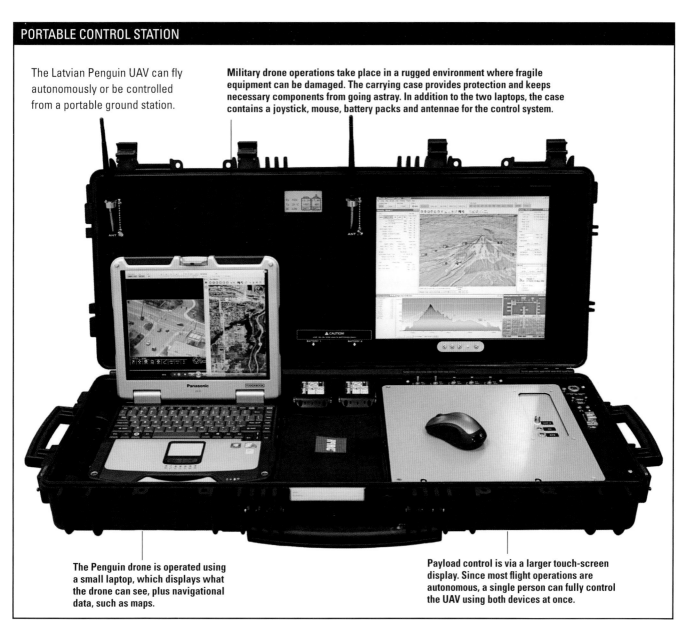

The Penguin drone is operated using a small laptop, which displays what the drone can see, plus navigational data, such as maps.

Payload control is via a larger touch-screen display. Since most flight operations are autonomous, a single person can fully control the UAV using both devices at once.

might mount a visual-imaging camera, as well as infrared and a laser rangefinder.

Lasers have many applications aboard drones. The military has used lasers for rangefinding for many years. A laser beam reflected from the target will be picked up by the drone's sensors and can be used to very precisely calculate the distance from drone to target. Of course, this information is only useful if the drone's location is also known, but GPS and other instruments will give the operator this information.

Laser rangefinding has many peaceful applications, not least altitude-finding. A rangefinder pointing down from the drone will tell it exactly how high above the ground it is (or how high above a point some distance ahead of its flight path, which may be a more useful piece of information!). Laser altimeters are more accurate than those based on air pressure, as they are not influenced by pressure changes due to temperature or weather effects. Not all drones need an altimeter, however. GPS positioning and a good map programmed into the control system can avoid ground hazards just as effectively, but by entirely different means.

Get-Home Plan

A drone using GPS and a programmed map is rather like a blindfolded person being guided around a room by someone calling out instructions based on a floor plan rather than direct observation. If all goes well, i.e. the plan shows all hazards accurately, there should be no problems at all. However, the plan will not show that the coffee table has been moved or that the family dog has decided to wander through the room.

The operator cannot know if something has changed since the map was created, and, of course, if the GPS signal is lost then the drone has no way of telling where it is or what is around it. Some drone systems have a 'get-home' navigational plan running while they are in operation. If contact is lost, then this plan is implemented and the drone flies home 'blind', based on the last data it obtained.

The use of instruments such as a laser altimeter allows a drone more choices when it is operating autonomously. Using data collected by the drone itself requires more onboard processing power but does

Left: A gimballed forward looking infrared (FLIR) system on a helicopter mount. Some military rotorcraft UAVs are essentially pilotless helicopters, and thus are quite capable of carrying systems designed for piloted versions of the same aircraft. Custom systems are needed for smaller drones with a lower payload capacity.

Above: The Mantis UAV is a technology demonstrator and testbed for combat UAV systems. It is designed to fill the same niche as the MQ-9 Reaper; a long-range combat-capable unmanned vehicle. Powered by twin turboprop engines, Mantis was designed for an endurance of over 24 hours' flight time.

Left: Based on Global Hawk, the MQ-4C Triton UAV is designed for long-range intelligence gathering and maritime surveillance. Among its potential missions are force protection and convoy escort, acting as a remote sensor platform to warn of threats that have not yet been detected by surface ships.

allow the drone to actively avoid hazards or choose a suitable route without needing to be programmed in detail.

Laser rangefinders can also be used to create very accurate maps, including mapping areas that are not accessible by other means. Drones have been used to map mountains down to millimetre scales, an undertaking that would not be possible, or at least not economically feasible, using other means such as conventional aerial photography or ground-based measurement.

Some military drones can be used to facilitate extremely accurate 'standoff' attacks, i.e. attacks launched with artillery and missiles from a great distance. One method is to use the drone's position and data from its rangefinder to obtain a precise target location, and then use this to programme GPS-guided weapons. However, this is of limited use against moving targets.

Laser Designator Systems

Some drones can carry a laser designator. Laser designator systems have been in use for decades now, but they have tended to be large devices carried by aircraft or infantry personnel. A designator small enough to fit in a drone but powerful enough to be useful was something of a technical challenge, but its benefits are considerable. Laser-guided weapons home in on a spot of laser light reflected from the target, and do so with great precision. They do not follow the beam, and therefore can approach the target from any direction.

The use of a drone-mounted designator allows a small, possibly

undetected drone, to point out targets for artillery shells, bombs or missiles launched from a distant platform. The target point can be moved to another target or can follow one that is moving. It can also be pulled off to send the weapons away from the target if some reason suddenly emerges why it should not be attacked. This allows extremely precise weapons, such as aircraft-dropped Paveway bombs, Hellfire missiles and the like, to be used against small or moving targets with great accuracy, and retains a 'man-in-the-loop' right to the very moment of impact, enabling the strike to be retasked or aborted in a manner that is simply not possible when launching conventional bombs, shells or missiles.

Using a designator in this manner allows operators to get the absolute most out of a missile salvo or group of shells. If the first missile strikes home, the aim point can be switched to the next most valuable target. If for some reason the weapon does not destroy its target, additional munitions can be used until it is hit, after which the remaining weapons can be switched to other targets. Without

this capability, it would be necessary to either launch multiple weapons at each target or risk not destroying it. By the time a second strike is made, the target might have left the area.

Traditionally, using designators in this manner has required infantry or vehicle-mounted observers to be close to the target, or the use of helicopters or aircraft. These platforms can be detected and attacked, while even a concealed infantry observer is close to the enemy and may be at risk. A drone is relatively inexpensive and may not be deterred at all, and there is an additional dimension to this.

Hostiles who spot a drone cannot know if it has a designator aboard, and if so, whether or not there are already weapons in the air headed for their position. A drone could be used as a very effective deterrent in this manner, acting as a broad hint that potential hostiles should remain concealed and under cover rather than launching an attack. It is not possible to say how many lives are saved by deterrence, but the obvious presence of a drone escorting a convoy might be the deciding factor that prevents

Above: The Watchkeeper UAV is in service with the British Army for surveillance and reconnaissance missions as well as target acquisition for artillery formations. A limited deployment was made to Afghanistan, where the ability of Watchkeeper's sensors to spot hostiles moving under cover of dust storms was found highly useful.

Left: The Talarion UAV is manufactured by EADS (European Aeronautic Defence and Space) for operations over land and sea, and also in the complex littoral zone between the two. France, Germany and Spain have all ordered the system in modest numbers. The twin-engine design offers enhanced 'get-home' capability in the event of damage or a system failure.

an attack. If not, then it does allow a rapid and accurate response to be made without risking the safety of friendlies or neutrals in close proximity.

Other Electronic Systems

Other systems carried by drones include specialist sensors (for example, for weather monitoring) and radio equipment used for purposes other than communicating with the ground station. A drone can act as a radio relay beacon, picking up transmissions and rebroadcasting them. This has many applications, including ensuring that military teams operating in a remote area have good communications with base, or helping responders in a disaster area retain contact with their supports.

It is even possible for specialist drones to act much like communications satellites. Rather than launching a satellite to 36,000km (22,369 miles) in a geosynchronous orbit, it is possible to use a high-flying drone to bounce signals over the horizon, which is not only cheaper, but also cuts out some of the delay imposed by the long distance that signals must travel. Interference may also be reduced by this method.

Not all signals that are picked up are retransmitted with the sender's consent, of course. Drones can carry out radio interception missions, flying above enemy territory to monitor their radio signals. At the very least this allows the operators to build up a picture of who is transmitting, how often and on what frequencies, which can allow intelligence analysts to determine strategic

information about the enemy, even if signals cannot be interpreted.

It is often possible to obtain information from decrypted signals, or those made 'in clear' by operators who think that they cannot be intercepted. This sort of signals-intercept work has long been the task of aircraft or concealed ground-based listening posts, but can now be undertaken more safely and at less cost. Just as importantly, it is possible to send drones into an area at short notice, enabling signals-intercept coverage to be set up very quickly and often covertly.

Other emissions can also be detected, such as radar. Again, much can be learned from the type and strength of radar emissions that are detected. For example, short-range air defence radar has very different characteristics to the navigational radar systems used by ships. Radar intercepts can be used to locate previously unsuspected enemy forces or air defence positions, enabling friendly aircraft or ground forces to be routed around them, or to launch an attack on favourable terms.

Radar and radio can also be jammed by drones, although this requires a great deal of power and is not possible with smaller drones. The electronic-warfare mission has traditionally been the province of specially equipped aircraft or those carrying a self-contained systems pod that gives some basic capabilities. A drone can now be sent ahead of an airstrike to provide electronic warfare capability, or a drone could be used to deceive enemy defences by being sent into an area to jam radar if a strike was en route. The resulting alert will distract attention and absorb resources while perhaps lulling the enemy into complacency with a series of false alarms.

These missions can be undertaken by other platforms, but, again, drones are cheaper and can thus offer more coverage for the same investment. A ground-based electronic warfare

formation can use drones to widen its area of operations or to conceal its location, using the drones to transmit powerful signals rather than relying on the antennae at the unit's main location.

Similarly, a drone can be used as a radar platform. One significant problem facing drone manufacturers was how to squeeze a radar set into a small vehicle and how to power it. Generating radar pulses requires a great deal of energy that can only be supplied by a drone that has a sufficiently large engine to generate power for the radar; it is not possible to put a battery-powered radar set aboard a small drone.

SAR

Synthetic Aperture Radar (SAR) is used aboard some drones as well as aircraft. This system uses a fixed emitter pointing sideways out from the vehicle and makes use of the movement of the aircraft. A series of radar pulses is sent out as the aircraft or UAV moves. These are reflected from objects they encounter and return to a detector carried as part of the radar set. One pulse and its reflections will produce a very limited picture of a situation, but over time the images produced by successive pulses can be built up into a detailed model of the area being scanned.

SAR imaging is highly useful for mapping and for detecting fairly slow-moving objects, but it is of very limited use for combat purposes. Thus a SAR-equipped drone can be used to make maps, or to build up a picture of shipping around the parent vessel – or for search and rescue purposes. It is not capable of tracking a fast-moving incoming aircraft, or providing a firing solution for defensive guns or missiles.

Radar of this sort is 'active' in military parlance, i.e. it does not simply receive data, but sends out signals and detects what comes back. A camera or thermal imager is 'passive', in that it detects heat or light that does not originate from the sensor device. Cameras might use sophisticated electronics to enhance low-light images or to translate thermal radiation into visible displays, but they only make use of what is there. Passive sensor systems are relatively covert and do not require much energy, but the use of active systems takes up power and can be detected by other sensor systems.

Active radar is used for applications from mapping and navigation to targeting weapons, but has the drawback that radar signals can be intercepted at a great distance. The effect is somewhat like driving along a country road at night. Without headlights, the driver may be all but blind and has little chance of spotting hazards or even staying on the road, but his lights can be spotted at greater distances than they are useful to him, which is a drawback if he wishes to remain unobserved.

For civilian drones carrying radar for mapping or navigation purposes, this is not a problem. Military drones risk detection when they emit radar signals, just as with any other emission, such as radio. However, the ability to put a radar-equipped drone up is very useful. It can be used to widen a search when seeking survivors of a disaster, or to increase radar coverage to protect a naval task force from attack. The drone might be attacked by anti-radar weapons, but this in turn provides protection to the main platforms – it is better that a drone is shot down than a ship sunk, in terms of cost and also loss of life.

Below: Synthetic Aperture Radar (SAR) units can be fitted to a great variety of aircraft. In addition to military applications, SAR systems can be used for terrain mapping, oceanography, meteorology and also to assist in rescue or disaster relief operations. SAR systems have even been used to look for water on the moon.

SYNTHETIC APERTURE RADAR (SAR)

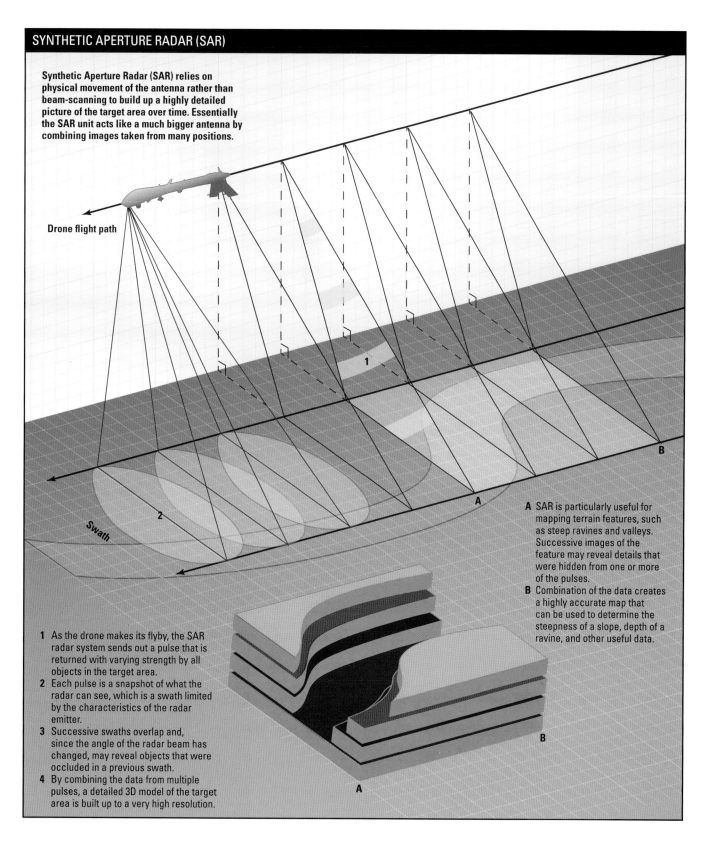

Synthetic Aperture Radar (SAR) relies on physical movement of the antenna rather than beam-scanning to build up a highly detailed picture of the target area over time. Essentially the SAR unit acts like a much bigger antenna by combining images taken from many positions.

Drone flight path

Swath

1

2

A

B

A SAR is particularly useful for mapping terrain features, such as steep ravines and valleys. Successive images of the feature may reveal details that were hidden from one or more of the pulses.

B Combination of the data creates a highly accurate map that can be used to determine the steepness of a slope, depth of a ravine, and other useful data.

1 As the drone makes its flyby, the SAR radar system sends out a pulse that is returned with varying strength by all objects in the target area.

2 Each pulse is a snapshot of what the radar can see, which is a swath limited by the characteristics of the radar emitter.

3 Successive swaths overlap and, since the angle of the radar beam has changed, may reveal objects that were occluded in a previous swath.

4 By combining the data from multiple pulses, a detailed 3D model of the target area is built up to a very high resolution.

Above: This composite radar image of Haiti was taken from high altitude by a NASA UAV. Data is collected in blocks corresponding to a target area, then combined to create a larger or more detailed image. Since the SAR unit has to fly back and forth, imaging a large area can take some time.

SEAD

Conversely, a drone can be used to detect radar emissions without revealing its location. Passive radar, as this is known, makes no emissions, but can pick up signals – much like someone watching for the headlights of a car on a dark road might not be spotted at all, but will almost certainly see the bright lights. Passive radar can be used for intelligence purposes as already noted, and can also be used to launch a pre-emptive strike on enemy radar equipment.

This technique can be used as part of SEAD (Suppression of Enemy Air Defence) operations. A platform (traditionally an aircraft, but a drone could also be used) equipped with missiles is

slipped into enemy airspace as stealthily as possible, and builds up a map of enemy air defence radar emissions. When the main strike comes into range, enemy radars are switched on to track the incoming aircraft, little knowing that a platform armed with anti-radar missiles is right on top of their position already.

These are not new capabilities, but new applications have been made possible by mounting miniaturized systems aboard drone vehicles. The small size and general stealth of drones makes them useful intelligence gathering platforms, and enables techniques like prepositioning drones over an enemy-held area to launch strikes when hostiles reveal their presence by turning on their radar.

The ability of drones to carry so many different information gathering systems is highly useful, especially since they can do so without exposing people to risk and at relatively low cost. A drone can be risked to find out a piece of information that would not be worth losing lives over, or can be sent on a one-way mission. Reduced costs compared to aircraft and helicopters mean greater coverage for the same budget and some coverage where previously it was not affordable.

There has been no 'drone revolution', but instead a gradual widening of capabilities that is bound to continue as equipment like cameras and radar packages become cheaper and lighter, and power systems increase in power and duration. Already some impressive capabilities are available, and, now that the technology has been proven, it seems highly likely that this trend will not merely continue, but will accelerate.

SUPPRESSION OF ENEMY AIR DEFENCES (SEAD)

Tracking and anti-aircraft systems can be identified by their radar emissions, as each application requires radar with different characteristics. This does not require the drone itself to emit any signals; it merely 'listens' with its passive radar receiver. Once enemy systems are detected and located, they can be attacked and put out of action.

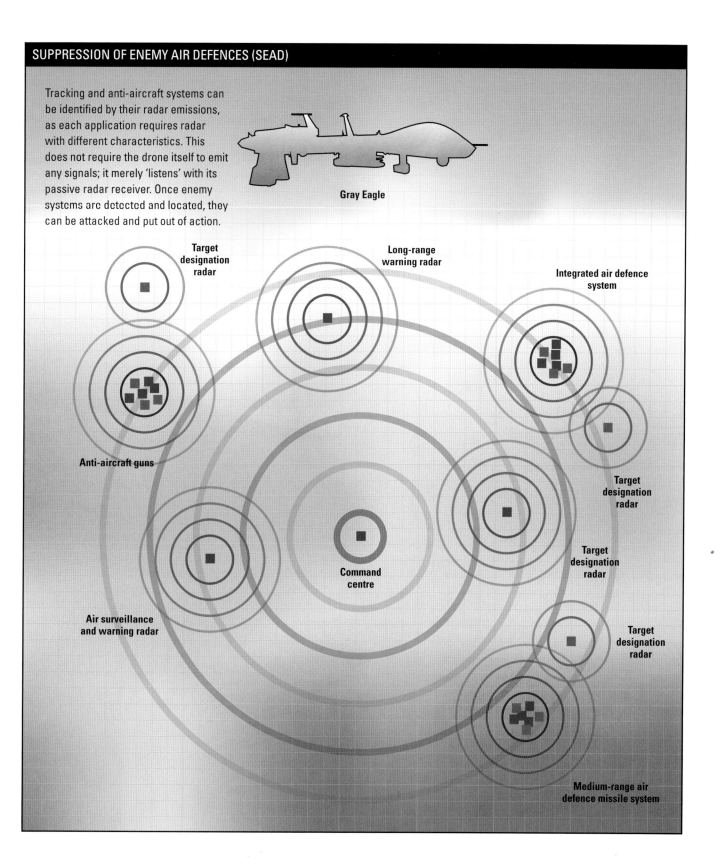

Gray Eagle

Target designation radar

Long-range warning radar

Integrated air defence system

Anti-aircraft guns

Target designation radar

Air surveillance and warning radar

Command centre

Target designation radar

Target designation radar

Medium-range air defence missile system

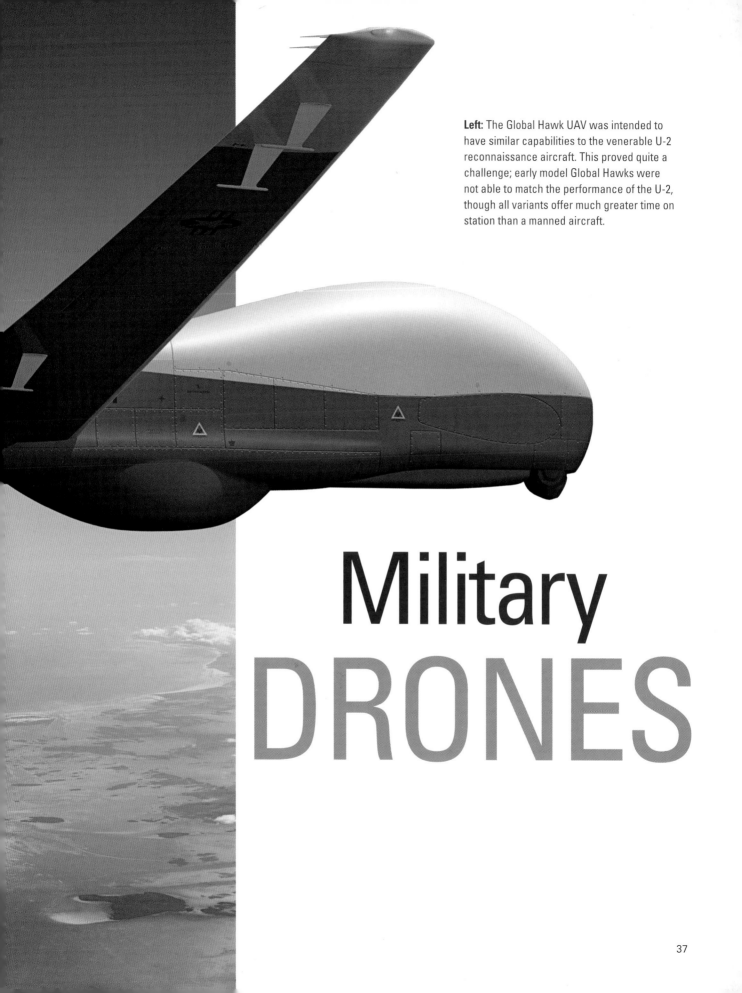

Left: The Global Hawk UAV was intended to have similar capabilities to the venerable U-2 reconnaissance aircraft. This proved quite a challenge; early model Global Hawks were not able to match the performance of the U-2, though all variants offer much greater time on station than a manned aircraft.

Military
DRONES

Military drones

Modern warfare has been described as taking place in a four-dimensional battlespace rather than on the traditional battlefield. Although a rather strange sounding concept when first encountered, the idea of a 'battlespace' grew out of changing circumstances that made the traditional model obsolete. A ground or sea battle can be influenced by electronic systems and weapons carried by air assets, forcing commanders to think in terms of airspace and the electromagnetic spectrum in addition to terrain and weather.

Above: This World War I German observation balloon effectively created artificial high ground for battlefield observation. Although primitive, balloons proved useful, although they were, by their very nature, static and rather vulnerable. Powered aircraft made balloons obsolete, not only by taking over their role, but also by providing an effective means of destroying them.

Before the development of aircraft, it was relatively simple to establish a front line and a fairly safe rear area. Enemy forces could infiltrate or perhaps make a long flanking movement to get into the rear, but otherwise the only way to reach targets behind the front lines was to fire artillery shells at points marked on a map and hope for the best.

The advent of reliable motorized ground transport changed this situation to a significant degree, as it became possible to launch deep-penetration attacks designed to break through a line and cause havoc in the rear areas. Motorized mobility also allowed a rapid shift in the axis of attack, creating a more fluid battlefield situation. However, ground forces still had to be supported, fed and refuelled, and this reliance on supply lines meant that fast-moving mobile formations were still tied to a system of logistics bases. If cut off, they would quickly run out of fuel and ammunition and be overrun.

The use of aircraft for combat and support missions opened up a wider range of possibilities. Although still tied to bases, aircraft could reach areas far behind the battle lines or, indeed, very distant from the combat zone and strike targets before returning home. It became possible to obtain detailed reconnaissance about not merely what was on 'the other side of the hill' but what was going on deep in enemy territory.

Three-Dimensional Battlefield

By World War II, the situation was far more fluid than it had previously been and the simple linear battlefield model no longer sufficed. A combination of air strikes and parachute drops or a fast ground advance could seize areas that might have seemed safe. Aircraft could strike at rear-area command and communications sites, supply dumps, troops moving to and from the battle area, and the infrastructure that supported the whole endeavour. The battle area became deeper and also wider, with fewer guarantees about which areas were safe and which areas might be contested.

This created a large three-dimensional battle area in which control of the air, or at least denying airspace to the enemy, was of paramount importance. To these dimensions was added the fourth – the electromagnetic spectrum. Electromagnetic effects include radio and radar, infrared detection and thermal imaging, conventional and low-light cameras and, of course, various ways of denying the use of the EM spectrum to the enemy, such as jamming radio signals or attacking radar sites.

Improved reconnaissance, that has perhaps been gathered as simply as by having an observer radio what he sees from an aircraft, has enormous implications for the personnel on the ground. Until fairly recently, this capability was largely restricted to organized military forces that had the necessary resources. Changing technology has altered that situation, however.

In today's world, a significant degree of capability can be obtained without any military equipment at all. Binoculars with a built-in laser rangefinder are commercially available, enabling an observer to accurately determine the distance to a target from his position – which he can pinpoint using the GPS receiver on his inexpensive mobile phone.

THREE-DIMENSIONAL BATTLEFIELD

To drive the enemy from his positions, ground forces must advance and engage hostiles directly. Visibility is limited at ground level.

Airborne assets can provide an overview of the battlefield, feeding information to the ground forces and receiving requests for support.

Air power rarely achieves decisive results alone, but air support can move quickly to where it is needed without regard to terrain or enemy forces in the way. There are no 'front lines' in the sky.

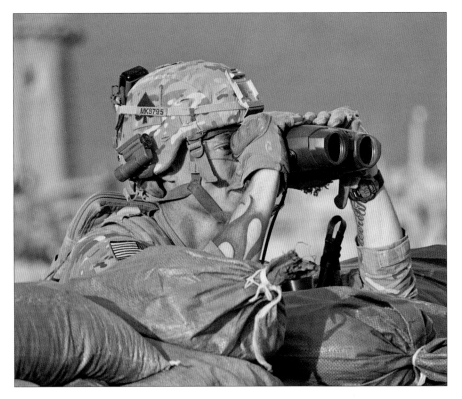

Left: Rangefinder binoculars use a built-in laser to provide an extremely accurate measurement of distance to whatever the user is looking at. Electronic systems of this sort have become commonplace and are well integrated into the combat information network. Similar devices are available on the open market, making these capabilities available to anyone.

The latter can also be used to trigger an explosive device, or inform allies of the target's position and movements.

The observer might, depending on the circumstances, not even have to conceal himself or his actions. Western militaries often operate under strict rules of engagement, especially where there is no declared war ongoing, and may not be permitted to act against someone who is reporting their actions even if it is obvious that he is supporting hostile personnel.

Tank Plinking

If military equipment is available, then the possibilities increase considerably. During the 1991 Gulf War, Iraqi tanks parked at night acquired a disturbing tendency to mysteriously explode. Some crews had been in the habit of sleeping in or under their tanks for warmth, but soon the rule of thumb became to get as far away as possible from your tank before bedding down for the night.

What was happening was that although the vehicle was well concealed from the naked eye by the darkness, it would be a different temperature to the surrounding terrain and would stand out on infrared or thermal imaging. Reconnaissance aircraft could see a cooling tank long after dark, and soon the practice of 'tank plinking' was born. Tank plinking was the term given to using any excessively heavy weapon to destroy a tank, such as a laser-guided bomb.

The tank crew had no way of knowing if there was a reconnaissance platform within range to see their tank, and, if so, whether or not a laser designator was

Above: Many of the Iraqi tanks destroyed in the 1991 Gulf War were 'plinked' with guided bombs after being detected using airborne thermal imaging equipment. Darkness no longer offers much cover, especially since UAVs can observe a battle area for long periods at a time with sophisticated sensors.

pointed at it. A bomb could be 'tossed' from a distant aircraft by climbing sharply, effectively launching it in an upward arc to extend its range. This would be extremely inaccurate with unguided munitions – there would be little to no chance of striking a single tank – but with a laser to home in on, or GPS guidance, the bomb would silently glide to the target and score a direct hit, or land close enough to annihilate an armoured vehicle.

Iraqi tank crews did not know why their tanks were exploding, but word got around very quickly that it was happening, with serious damage to morale. The psychological effect would have been similar, even if they did know why it was happening – the thought that at any time you might be targeted by a silent strike out of nowhere is disturbing to say the least.

Infrared detection is passive, i.e. there are no emissions from the detector to warn the target that he is being observed. A force that thinks it is moving quietly and under cover of darkness might be detected and tracked using a thermal

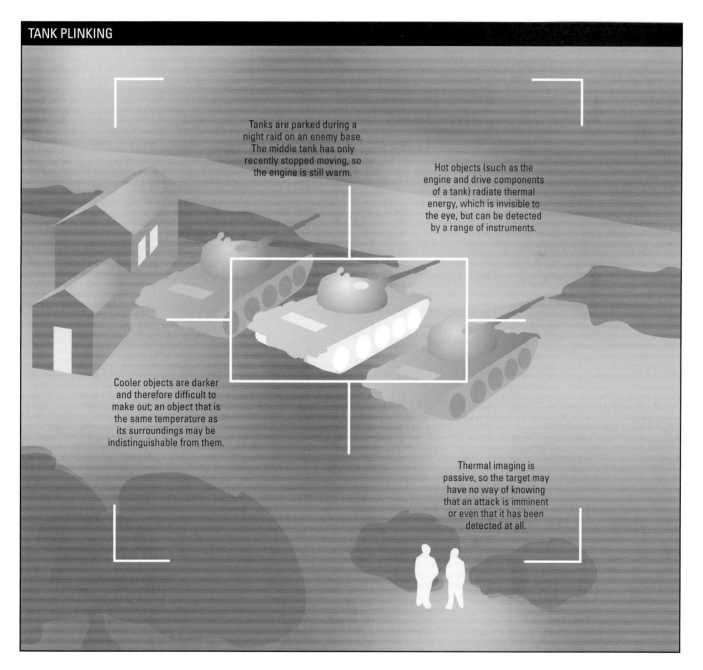

TANK PLINKING

Tanks are parked during a night raid on an enemy base. The middle tank has only recently stopped moving, so the engine is still warm.

Hot objects (such as the engine and drive components of a tank) radiate thermal energy, which is invisible to the eye, but can be detected by a range of instruments.

Cooler objects are darker and therefore difficult to make out; an object that is the same temperature as its surroundings may be indistinguishable from them.

Thermal imaging is passive, so the target may have no way of knowing that an attack is imminent or even that it has been detected at all.

camera without realizing it. Likewise, emissions from radar and radio systems can give away a unit's position to an unsuspected detector.

Thus the modern commander of a ground force must be mindful of more than what he can see or what is within rifle range of his patrol. Use of the electromagnetic spectrum enables a small force on the ground to bring to bear enormous firepower, and to do so very suddenly. An infantry unit may be warned that it is in proximity to the enemy in time to avoid contact or to stage an ambush, or can call in support ranging from artillery to air units.

The ability to call in a precise location and to guide weapons to it with great accuracy allows support of ground troops with quite heavy weapons, even against targets in close proximity. Traditionally, artillery needed to fire ranging shots and be corrected by ground observers, or else had to just bombard a map grid. Today, however, a shell can be guided to a target by a designator or to a spot on the ground by using GPS with a high degree of confidence. Thus the enemy

will receive no warning before bombs, rockets, missiles or artillery shells begin to land. The enemy force may never even become aware of the infantry formation calling in the strike.

The Modern Battlespace

All of this requires the use of the fourth dimension of modern combat, the electromagnetic spectrum. Attempts to deny its use have been made since the earliest uses of radio and radar. The simplest form of denial is to jam the signal by swamping it with a more powerful one on the same frequency. This is a brute-force method, but it can and does work, although modern radio and radar equipment is increasingly jam-resistant. Jammers also advertise their presence rather strongly, and many missiles have a 'home-on-jam' mode to remove them from the equation.

The modern battlespace is a complex and cluttered environment, made even more so by ambiguity about exactly who is a combatant and who is not. Direct conflicts between two uniformed military forces, with clearly defined combat

zones, are rare in the modern world. More commonly a conflict takes the form of low-level harassment by insurgents, planting of improvised explosive devices (IEDs) and guerrilla warfare by forces that will hide among the civilian population when not actively engaged.

In this environment, information is more important than ever. If a hostile group can be constantly tracked, then forces on the ground can be sure of its identity. Enemy bases can be found and identified by following the movements of combat or logistics formations. Intelligence gathered from a variety of sources and collated into a useful whole enables ground forces to make a faster and more accurate assessment of a developing situation than would otherwise be possible.

One highly useful factor here is obtaining information that the enemy does not think is available. A group of insurgents will act like ordinary people going about their daily business if they think they are being observed, but those that are not aware of observation are more likely to display weapons or behaviour identifying them as belligerents. Overt surveillance can sometimes be used as a deterrent, but covert information gathering is more effective in finding targets.

Thermal or low-light cameras, or detectors that indicate when certain emissions are present, often allow a group's activities to be monitored without their knowledge. This has numerous implications. A suspect group in a remote area might just be a roving band of tribesmen or even goatherds. Sending

Right: Thermal imaging can be difficult to interpret without training or practice. Shapes that would be readily identifiable under visible light can sometimes look quite different when viewed in terms of heat distribution. Depending on the type of imager, hot objects may be displayed as brighter or darker than cooler ones.

Below: Improvised Explosive Devices (IEDs) are a grave threat to ground forces. UAVs can help mitigate this threat by spotting disturbed areas where an IED may have been planted or observing insurgents in the act of emplacing one. They can also help protect personnel dealing with an IED by warning of hostiles in the area.

Above: Hand-held thermal imagers have a range of applications, including rescue work and damage assessment. A thermal camera can see what the naked eye cannot, and can often penetrate visual obscuration or camouflage. It can also provide useful information, such as whether a vehicle has been started or driven recently.

ground forces to investigate them might cause ill-feeling if they are neutral or even friendly, which can worsen a difficult political situation. It also absorbs manpower and exposes the force to the risk of ambush or an accident en route.

A period of covert observation might uncover the truth about the activities of the suspect group – who might, after all, be simply looking after their goats! However, this again requires getting an observer into the area. A ground team will take a long time to reach a remote spot; aircraft and helicopters may be detected. Once the suspects have realized they

are under observation, they will likely move off or conceal their activities. The ideal outcome is a covert observation that includes multiple detection systems. The suspect group may do a great job of making sure that they really do resemble goatherds visually, but emissions from their air defence radar system or regular radio traffic from their position might tell a rather different story.

Avoiding Collateral Casualties

There is another dimension to modern warfare: the manipulation of public opinion and the difficulty of obtaining a true

picture of events. It is not uncommon for both sides in a local conflict to blame one another for tragedies or atrocities, and staging of incidents is also a commonly used gambit. Images of innocent people killed in an air or artillery strike are a powerful way of influencing the opinion of the voting public in Western countries, but so is footage of the 'victims' getting up and congratulating one another on a well-staged scene after their cameras have shut down.

Tragedies, whether staged or real, are one way to influence foreign opinion. Sympathy among the voting public or demands for the cessation of air strikes can influence the strategy of democratic nations, and most modern commanders understand this. Dramatic footage of wrecked military vehicles can give an

HOW LASER-GUIDED MISSILES WORK

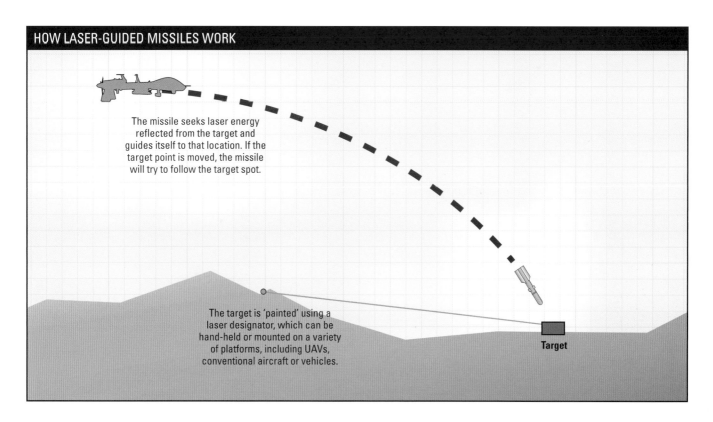

The missile seeks laser energy reflected from the target and guides itself to that location. If the target point is moved, the missile will try to follow the target spot.

The target is 'painted' using a laser designator, which can be hand-held or mounted on a variety of platforms, including UAVs, conventional aircraft or vehicles.

Target

Right: Reconnaissance by UAVs helps find arms caches, such as the one being destroyed here, and targeting data can be provided to enable a 'standoff' strike using precision-guided weapons. This gives insurgents no time to remove their weapons and hide them somewhere else and does not expose personnel to much risk.

impression of a conflict that is being lost, which again can lead to calls for a withdrawal. It does not matter to an insurgent group whether it defeats its enemies or obtains their withdrawal – or forces them to the negotiating table by way of pressure from foreign countries. Whatever tools are used to win a conflict, it is the outcome that matters.

For personnel operating in this complex four-dimensional battlespace, there is a constant need to be sure of targets, to be restrained in the use of force, even when under fire, and to avoid collateral casualties, even when the enemy is firing from among his own

Above: If the decision to launch a missile is taken by a human who is somewhere nearby – in this case the pilot of an F-16 – then this is an accepted part of warfare, even though the weapon may actually guide itself to the target. Using a UAV to make the same strike is more controversial.

civilians. This creates circumstances where many weapons cannot be used.

Unguided mortar bombs and shells may not be precise enough to satisfy a public that has become used to hearing of 'surgical strikes' and 'precision-guided weapons', and even many guided weapons can be problematic. A missile that homes in on its target by radar or thermal means will hit that target with a high degree of precision, but that does not mean it will not cause collateral casualties if civilians wander into the

blast radius, or if the target is suddenly identified as non-hostile.

The answer put forward by some manufacturers of guided weapons is to use optically guided missiles, with a camera onboard that feeds back to a human operator. Up to the last second there is an option to take the weapon off target or to retask it. This arguably enhances precision, but more importantly allows the attack to be aborted.

This concept has been recognized by arms manufacturers for some time – and

has caused some controversy, since the idea is not universally accepted. It also applies to drone operations. There have been questions about the legality or morality of arming unmanned vehicles. The question revolves around whether it is ethical to use drones for direct combat action, and there are differing opinions on the subject.

Drone Weapons

Some opponents of drone weapons seek to portray them as some kind of robotic death machine with the capability to maraud across the landscape attacking anything that appeals to its twisted programming. It would seem unwise to create such a 'war robot' with complete autonomy. Fortunately, this scenario is quite far from the reality of the situation.

Automated weapons have been in use for some time. A missile or torpedo, depending on its guidance system, can be locked on target and sent on its way. At that point it is beyond human control (as already noted, some manufacturers strongly advocate 'man-in-the-loop' operations to allow an attack to be aborted if necessary, but this is not usually the case) and will carry out its destructive mission. Some air defence weapons are also automated and will fire on targets that do not give the correct IFF (Identification Friend-or-Foe) response, and that meet their pre-programmed target parameters.

Yet these are not rampaging death machines, and they have been an accepted part of warfare for some time. A human makes the decision to launch a missile or activate the air defence system, and can take it offline or make a no-shoot

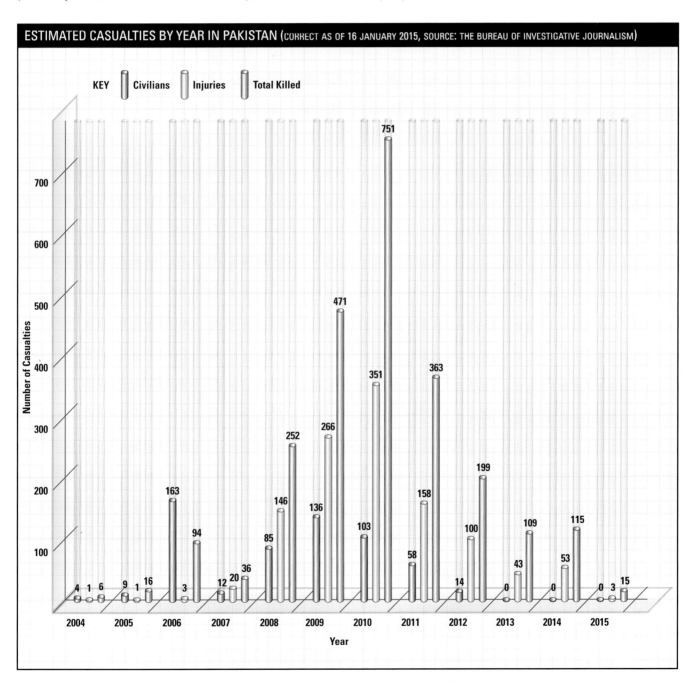

ESTIMATED CASUALTIES BY YEAR IN PAKISTAN (CORRECT AS OF 16 JANUARY 2015, SOURCE: THE BUREAU OF INVESTIGATIVE JOURNALISM)

KEY: Civilians | Injuries | Total Killed

Above: Precision-guided weapons can be used to strike hardened targets, even when they are in close proximity to civilians. Several of Saddam Hussein's critical buildings were attacked in this manner during 2003. The public has become so used to this level of precision that questions are asked when it is not achieved.

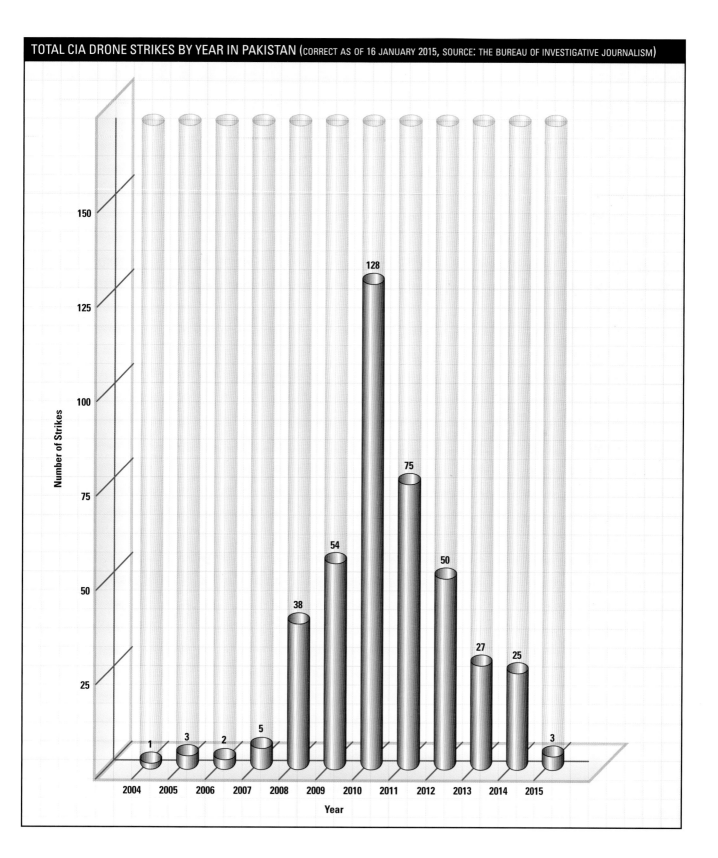

TOTAL CIA DRONE STRIKES BY YEAR IN PAKISTAN (CORRECT AS OF 16 JANUARY 2015, SOURCE: THE BUREAU OF INVESTIGATIVE JOURNALISM)

Above: Despite its menacing name, the MQ-1 Predator UAV was developed first and foremost as a reconnaissance platform. The capability to carry weapons was added later and required significant development work. The US military is quite open about reconnaissance operations with the Predator, but considers combat operations classified information.

decision at need. Similarly, all drones that carry weapons such as missiles or bombs are operated by humans who have input into the targeting process. A small number of drones are weapons themselves, i.e. they function in a manner similar to a missile, and morally these are little different to a guided missile of any other sort.

Ethically, there seems to be no real difference between a drone operator making the decision to launch a weapon and a member of the crew of an aircraft or a ship located offshore doing the same. An artillery commander has the same decision to make and is equally removed from the impact location. One of the moral objections to drone

weapons is the idea that a distant operator may not really grasp the import of what he is about to do, that it is somehow 'just like a video game' and therefore the decision to kill can be made too lightly.

This idea has been challenged on the grounds that drone operators know very well what they are doing and what its implications may be; just as artillerymen, naval gunners and strike aircraft crewmembers do. It is true that, generally, it is easier to do violence from a distance, but if that is the main

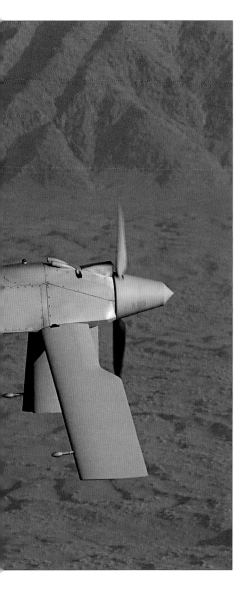

who orders forces into action, or the politicians who declare a war that is then fought by others.

Perhaps a more useful question is: how much autonomy is it ethical to give a drone weapon system? A human makes the decision to drop a bomb or fire a missile from a drone or aircraft at a given target; a human makes the decision to fire his rifle at a particular enemy soldier. Both situations seem to involve the decision to make a fairly direct attack on a particular target chosen by the attacker, and both allow a reasonable degree of surety that it is the chosen target that will be attacked.

On the other hand, sending out a drone with programming to attack whatever targets match a set of preset parameters might be straying into uncharted moral waters. The decision of what to attack and what to spare now rests with a machine. Human choice ends with the implementation of programming and the setting of target parameters. This means that there is no 'man-in-the-loop' at the moment the decision is made to kill someone. There are many that argue that this is a step too far. It may be efficient at times, but there are moral issues with sending out a machine with orders to find something suitable to destroy. Tight target selection parameters might make this acceptable, but there are many who believe that this removes humans too far from the decision to do violence.

Thus it is perhaps not physical distancing that makes arming drones a morally difficult question, but the number of critical decision points between the last human involvement and the effects on the target. If a missile is launched by a human – be it from a remotely controlled drone or some other launch platform – and guides itself to the target autonomously, then there are no decision points between a person and the act of violence. If a drone is programmed with targeting parameters and sent out on a

mission, spots a suitable target, evaluates it and decides to attack, then there is a decision point after the last human interaction. The decision to harm that particular target has not been made by a person, but by a machine.

It may be that the moral question of arming drones comes down to the concept that we owe it to our fellow humans – even our enemies – to take responsibility for the harm we do them. If a person has to make the choice to kill and feels that it is justified, this is within our existing moral framework. Giving that decision to a machine requires some new thinking that we have not yet properly begun to understand.

Of course, whatever the moral implications may be, the legal situation will be influenced by other factors as well. Necessity often drives some ethically grey decisions, and likewise morally justifiable actions may be made illegal for political reasons. The legalities of drone operations are only just beginning to be explored, but it seems likely that unless there is some powerful political reason for outlawing armed drones, they have proved to be sufficiently useful that they will remain in use.

The big question is going to be: how much autonomy are we willing to give to armed drones … and will that turn out to be a little too much?

Integrating Drones into the Modern Battlespace

The availability of reliable, lightweight computer and communications equipment has permitted the rise of 'network-centric warfare'. Under this model, information is shared in a peer-to-peer manner among the forces involved rather than data being passed 'up the chain' by each formation and then down again to those deemed to need it.

Communications between forces in different chains of command, especially during joint operations with different arms of service involved, have traditionally been

objection then, in the words of one military thinker: 'We started down a morally ambiguous road the moment somebody came up with the idea of throwing rocks'.

If we accept that violence has been done from increasing distances throughout human history, it is difficult to pinpoint exactly when it becomes morally objectionable to be distanced from the results. Is violence only acceptable if you can see what you have done? If so, drone operators may be on morally stronger ground than the commander

Above: A Predator drone taking off. Long-winged, piston-engined UAVs of this sort have a somewhat flimsy appearance and relatively low performance, but more recent jet-propelled UAVs can, in some cases, hurl themselves into the air and make a hard break in the manner of a jet fighter.

prone to weakness. Critical information from an army infantry patrol might take some time to reach the naval warships supporting them from offshore, leading to an opportunity being lost.

A notable example of this problem occurred during the 1991 Gulf War, when Iraqi mobile SCUD missile launchers were being sought. Ground teams at times reported launcher locations, but it was several hours before an air strike against them could be mounted. This meant that opportunities were lost, often

due to slowness of data transfer.

Under the network-centric warfare model, all forces in a region are tied into a data-sharing net and can pass information directly from one to another. At the command level, this allows a detailed picture – perhaps too detailed at times – to be built up of what is happening on the ground. Micromanagement of forces is tempting when the commander can see what they can; it still remains a poor choice, however.

The big advantage of network-centric warfare is the ability to effectively use the eyes and the sensors of other units. An infantry patrol cannot see what is on the other side of the building they are approaching, but a drone linked into their network can send real-time images of what is there. This reduces the chance of ambush and gives ground forces a significant advantage when using their own weapons. It also facilitates the employment of supporting weapons, such as artillery or air assets.

Networking gives ground forces advantages that may often be quite small but nevertheless can be significant. For example, if an enemy armoured vehicle suddenly emerges from concealment, the crew of a friendly tank have to spot it, acquire it as a target, make the decision to engage and then fire. This process is shortened somewhat if a different platform has warned the crew that the enemy vehicle is approaching.

The process is still shorter if the friendly tank commander has a live feed from a drone or other platform that is observing the enemy vehicle. He now knows exactly where it is and when it is about to emerge into his field of fire, and can have a shot set up for the instant the hostile vehicle presents itself. This gives the enemy no reaction time and creates a very significant advantage.

Drones fit into the network-centric model of warfare at various levels, depending on their size and capabilities. Long-duration reconnaissance drones, perhaps equipped with multiple sensor systems, can monitor the battlespace and the region around it, building up the 'big picture' using cameras, thermal imaging and radar. Closer to the combat area, smaller drones can be used for tactical missions that include post-strike reconnaissance after artillery or air weapons have been used, sniper location, or tactical reconnaissance of the general situation.

Network-centric warfare enables close cooperation between forces belonging to different services, providing the necessary communications equipment is in place. Often this is nothing more complex than a ruggedized version of a civilian laptop or tablet; protocols for communication between devices have been in place for years and are available for military use as well as commercial applications.

Traditionally, information and orders have been pushed down the chain of command to those actively engaged with

Below: During the 1991 Gulf War, Iraqi SCUD missiles were used against targets in Israel, threatening to draw Israel into the conflict with political consequences for the Coalition. Countermeasures included trying to find the mobile launchers and destroy them, and shooting down the missiles in flight. Neither was spectacularly successful.

Below: An M1 Abrams tank operating alone is a powerful asset, but as part of a network its capabilities are greatly enhanced. Reconnaissance data from aircraft or UAVs allows a tank commander to see what is 'on the other side of the hill' and prepare for it, bringing his tank's fighting power to bear as effectively as possible.

the enemy, although some militaries have long held the belief that it is those at the front who are best positioned to see what needs to be done. This is a somewhat different model of combat operations, with small-unit leaders 'pulling' the force as a whole forward. The theory behind this model is that a junior commander

on the ground can spot an opportunity and act upon it immediately, informing his superiors of what he is doing so that they can support his actions and send assistance as required.

Thus rather than originating with a distant commander, offensive actions and battlefield manoeuvres would be

Right: A mobile SCUD erector-launcher. Locating mobile launch units can be a problem, and information on their location is time-sensitive. If an attack cannot be made before the unit moves on, the opportunity is wasted. This is what happened on numerous occasions in the 1991 Gulf War.

generated at low level, with higher commanders retasking assets and feeding in additional units to support what is already taking place. This has been done with some success, although there is some suspicion of these methods among higher commanders who are used to having total control over the combat operations they are responsible for.

This 'pulled from the front' rather than 'pushed from the headquarters' model of warfare requires a high level of skill and initiative among junior commanders, and is only effective when carried out

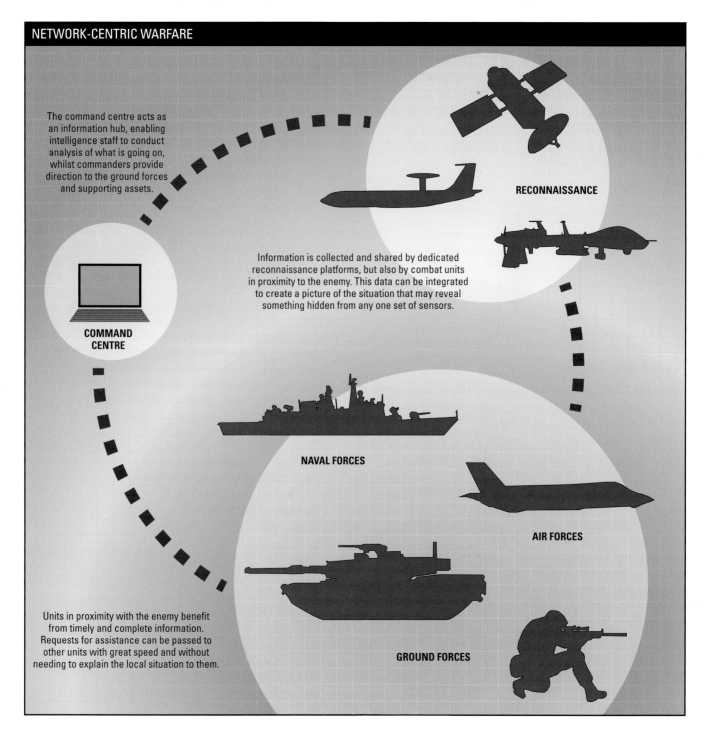

NETWORK-CENTRIC WARFARE

The command centre acts as an information hub, enabling intelligence staff to conduct analysis of what is going on, whilst commanders provide direction to the ground forces and supporting assets.

COMMAND CENTRE

Information is collected and shared by dedicated reconnaissance platforms, but also by combat units in proximity to the enemy. This data can be integrated to create a picture of the situation that may reveal something hidden from any one set of sensors.

RECONNAISSANCE

NAVAL FORCES

AIR FORCES

Units in proximity with the enemy benefit from timely and complete information. Requests for assistance can be passed to other units with great speed and without needing to explain the local situation to them.

GROUND FORCES

Above: The Switchblade drone is one of the few that is actually a weapon. Little more than a self-guiding warhead, Switchblade could be considered a missile or similar weapon, but is generally considered to be a drone. Weapons of this sort remain controversial, but do offer an additional capability to ground forces.

by a well-trained and cohesive force whose officers can think flexibly and react quickly. The move to network-centric warfare does not make the task any easier, but it does allow commanders to obtain necessary information from the network in a timely manner rather than waiting for it to be offered or made available from a central source.

Drones play an obvious part in this information-age model of warfare as reconnaissance platforms, but they have other applications, too. Drones can act as communication relays where contact would otherwise not be possible, and can bring specialist sensors to bear where a commander needs a closer look at something. Few infantry formations carry signals-detection equipment or a radar set with them, however useful these items might be at any given moment in time. A drone can be tasked to assist as needed and is far more likely to be available than a manned aircraft, which costs far more and must fly from its base to the combat zone.

Drones can, of course, also carry weapons, and can act as airborne support platforms if heavier aircraft are not available. While the number of weapons available is not great, a precise strike with a guided missile or bomb on an enemy bunker can save casualties, or prevent a movement from becoming bogged down while other means – say armoured vehicle support – is brought up to clear the obstacle.

Even where there is not a complex network of sensor platforms, drones can offer the troops on the ground a local advantage. Small drones can be carried in a backpack and launched by hand when necessary, enabling the commander on the ground to put a set of eyes high up enough to see what is going on around his force.

This can be used for reconnaissance purposes, e.g. a force might move into the area and use its drone to perform reconnaissance before moving on, or

Above: A Trident missile at the moment of launch. One possible role for long-duration UAVs is to observe for ballistic missile launch plumes, creating the maximum possible warning time – or reassurance that they are not being launched at all.

it can be used for more direct combat support. A drone can be used to find out where enemy forces are, without risking an unexpected contact, or can be used to locate enemy support positions, such as mortar batteries or artillery observers, in a situation where doing so manually might pose an undue risk to friendly personnel.

Drones in Naval Roles

Drones are also highly useful at sea and when protecting vehicle convoys. It is not always possible to provide constant

air surveillance for the latter; there are only so many aircraft and helicopters available. Experience in Iraq showed that air support was extremely useful in avoiding or dealing with ambushes, which were particularly common along the main supply routes used during the 2003 Gulf War. A small, inexpensive drone can be used to observe the route ahead and warn of possible attack, and will be cheap enough to be available in sufficient numbers to maintain near-constant coverage.

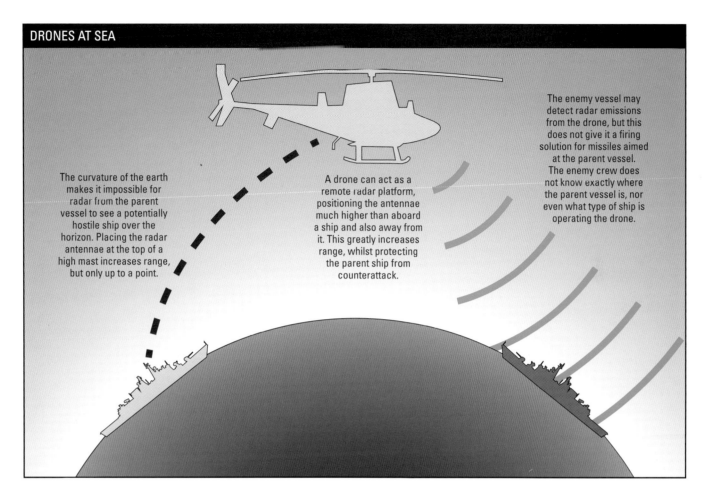

The curvature of the earth makes it impossible for radar from the parent vessel to see a potentially hostile ship over the horizon. Placing the radar antennae at the top of a high mast increases range, but only up to a point.

A drone can act as a remote radar platform, positioning the antennae much higher than aboard a ship and also away from it. This greatly increases range, whilst protecting the parent ship from counterattack.

The enemy vessel may detect radar emissions from the drone, but this does not give it a firing solution for missiles aimed at the parent vessel. The enemy crew does not know exactly where the parent vessel is, nor even what type of ship is operating the drone.

At sea, drones have the same advantages of small size and low cost. A warship can only afford so much deck space and so much topweight, and preference has to go to combat systems. Thus small warships, such as frigates, normally only carry a single helicopter. These are invaluable for search and rescue, missile guidance and anti-submarine work, but their capabilities are limited by the space available. Being smaller, drones can be carried in much larger numbers.

The most basic function for drones at sea will be the provision of reconnaissance. The sea is a huge place, and the horizon is really not very distant by comparison. A warship's radar can only see so far, which limits the warning time for an air or missile attack, and restricts the detection range for surface targets or anything else the warship is looking for. This might include a distressed vessel during a search and rescue operation – searches at sea level can take more time than the survivors actually have.

Thus drones at sea can be used to put a radar set or a passive radar detector high above a task force, increasing its range. They can also be used to position active radar away from the task force. This is important at times, since one of the commonest ways a naval force can be detected by hostiles is when it turns on its radar. Using a drone radar platform does not prevent detection, but it does give a false position.

Drones can also be used to search an area or to investigate a suspect ship, enabling a warship to 'eyeball' the target from distance without having to chase it. Even a very fast warship takes time to close from radar to visual range. An air vehicle can do so much more quickly, allowing the warship to move on to other tasks if the contact turns out to be benign. It also means that if the contact is hostile, the first target it fires upon is likely to be an unmanned and ultimately expendable drone.

Another use for aircraft and helicopters in naval warfare is mid-course guidance for missiles. Naval weapons have a long range, but need targeting data – you cannot shoot at what you do not know is there. One of the key principles of naval warfare is to 'Attack Effectively First', i.e. to hit the enemy and disable his vessels before he can hit you, or better still before

Above: Helicopters have proven eminently suitable for naval operations, since they can land on a small operating deck. The MQ-8 Fire Scout was based on an existing helicopter airframe, converted to autonomous operations. It can perform most of the functions of a manned helicopter, such as acting as a reconnaissance platform and radar picket.

he even knows you are there. Small, stealthy drones will fit into this model: they can find enemy vessels and guide missiles to them without being detected, giving the enemy no chance to respond. Even once he knows he is under attack, he may have no targeting data for a response and may be defeated without being able to retaliate.

Drones themselves may be able to launch attacks. The air groups operated by aircraft carriers, amphibious landing vessels and helicopter ships may be augmented or even replaced by armed strike drones, which does have the advantage of enabling carriers to be smaller, or drone air groups to be much larger. However, the ability to carry ship-killing missiles requires a larger drone, to the point where the tradeoff may not be worth it.

Dipping Sonar

Drones may become effective anti-submarine platforms in the future. One problem with using surface ships to hunt submarines is the need to take the ship within torpedo range of the target. This gives the advantage to whoever detects the other first, and, generally speaking, submarines have the advantage here. Helicopters are often used with 'dipping sonar' that is lowered from the hovering helicopter into the water.

Dipping sonar has the advantage that a submerged submarine cannot usually detect a helicopter, and may thus be unaware that it is being tracked. An attack can then be made by surprise, perhaps with torpedoes dropped from helicopters. Drones that can carry out these functions will need to be fairly large, but, since they would not need to also carry a crew, they could be smaller, cheaper and lighter. This would enable a small warship to carry a number of anti-submarine drones, although, as yet, the capability does not exist.

Drone combat aircraft have been mooted for a long time. In the 1970s, it was predicted with some confidence that the age of the manned fighter was over. Missiles and unmanned fighters were – at least in some quarters – expected to replace the conventional combat aircraft. There are advantages to a drone fighter, of course. It could be much smaller and cheaper than a manned fighter, and it would not be limited by the ability of the pilot or crew to withstand high g-forces during combat manoeuvres. However, as yet, drone fighters have still not emerged as such.

Some drones can carry air-to-air weapons, but these are adaptations of ground-based weapons intended for use against enemy helicopters and other less manoeuvrable targets. The appearance of high-g drone fighters is some way off, even if it is possible at all. There is also the question, as already discussed, of whether it is ethical to create such a machine. To be effective, a fighter drone might need to make its own combat decisions; ground control is possible, but many pilots agree that an operator who is not in the cockpit feeling what the aircraft is doing simply cannot be sufficiently effective in combat.

Thus, for the time being, the role of drones remains one of providing support and, mainly, reconnaissance data. However, they have already proven extremely successful in a great many applications and their role will surely widen. It remains to be seen whether drones will ever be a decisive, as opposed to a supporting, combat system, but one has already taken the surrender of enemy troops.

In 1991, after bombardment by US naval forces, Iraqi troops defending Failaka Island, in the Persian Gulf, surrendered to the first Coalition unit they saw. This was an RQ-2 Pioneer drone sent to perform post-strike reconnaissance. Incapable of carrying weapons and resembling an overgrown toy plane, this drone was nevertheless seen by enemy personnel as a way to have their surrender recognized and thus to avoid further naval bombardment. Its designers probably never envisaged such an occurrence, and it is likely that there will be a few more surprises as drone technology matures and becomes more prevalent in the modern battlespace.

DIPPING SONAR

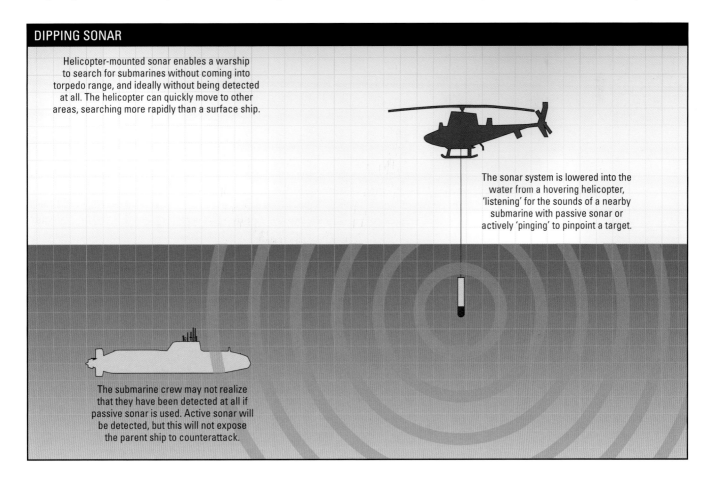

Helicopter-mounted sonar enables a warship to search for submarines without coming into torpedo range, and ideally without being detected at all. The helicopter can quickly move to other areas, searching more rapidly than a surface ship.

The sonar system is lowered into the water from a hovering helicopter, 'listening' for the sounds of a nearby submarine with passive sonar or actively 'pinging' to pinpoint a target.

The submarine crew may not realize that they have been detected at all if passive sonar is used. Active sonar will be detected, but this will not expose the parent ship to counterattack.

Military Drone Equipment and Weaponry

Some of the systems carried by drones have many civilian or commercial applications. Cameras in particular have many different roles and even quite

Left: Conventional helicopters equipped with dipping sonar are carried by many warships, including aircraft carriers. Drone helicopters can be smaller, enabling more to be carried in the same space or making room for other aircraft or UAVs, or can carry more fuel instead of crew, extending time on station.

specialist equipment, such as radar, can be used for mapping, environmental monitoring or search and rescue operations. Some systems are only useful in a military or strategic role, such as for covert information gathering. Among these are low-observable technologies, better known as 'stealth'.

The day may come when drones can carry devices that make them invisible, but current technology relies instead on making them difficult to detect. This is a significant difference; a drone cannot be 'cloaked' with existing technology and will

Above: Small drones, such as Puma, have a limited range and payload, but can still fulfil many roles. The ability for a warship to 'eyeball' a suspect vessel without having to close with it is very useful, as is the capability for an amphibious warfare vessel to observe the situation ashore.

be spotted by a person or detector that gets close enough. What low-observable technologies do is to reduce detectability so that detection distances are very short. The odds of passing that close to a detector or observer are quite low, so

SEARCH AND RESCUE

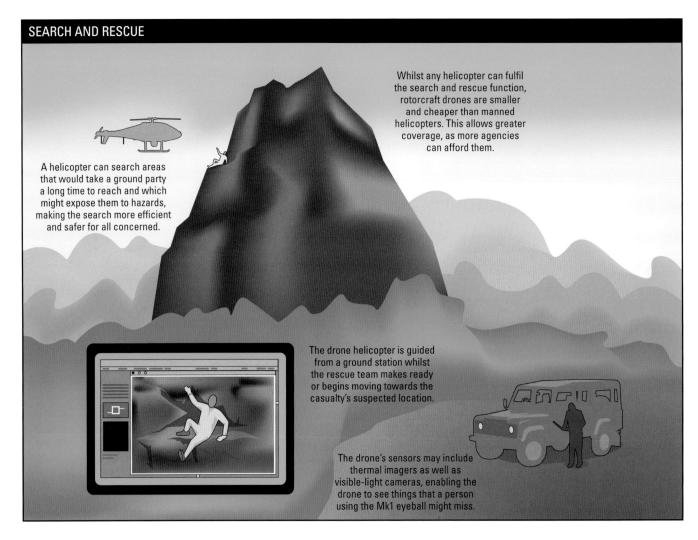

Whilst any helicopter can fulfil the search and rescue function, rotorcraft drones are smaller and cheaper than manned helicopters. This allows greater coverage, as more agencies can afford them.

A helicopter can search areas that would take a ground party a long time to reach and which might expose them to hazards, making the search more efficient and safer for all concerned.

The drone helicopter is guided from a ground station whilst the rescue team makes ready or begins moving towards the casualty's suspected location.

The drone's sensors may include thermal imagers as well as visible-light cameras, enabling the drone to see things that a person using the Mk1 eyeball might miss.

stealthy drones will be able to operate in areas where more obvious craft would be noticed almost immediately.

The detection of drones or other craft depends on various 'signatures'. The larger a signature is, the further away the craft can be detected and the more generally obvious it is. Signatures include visual, thermal, audible and electromagnetic, as well as radar cross-section.

A drone's visual signature dictates how likely it is to be spotted by the naked eye or a conventional camera. Small size is an advantage here – a large drone is more likely to be spotted than a very small one – as is colouration. A pale grey

drone is more likely to be missed against a cloudy sky than a black or bright red one, although which colours are the most 'stealthy' depends on the local conditions and whether the observer is looking up or down at the drone. The topsides of combat aircraft that are intended to fly low are often a different shade to the bottom, in the hope of blending in against the ground. Since many drones fly very low, it is quite possible that they will be seen by an observer located at a higher point.

Movement is also a factor. An object that is moving in a natural manner is more likely to be ignored than one that hovers, changes direction suddenly, or moves

in a jerky fashion, although hovering does allow a drone to conceal itself against certain backgrounds. Beyond a certain distance, it is very hard to spot a helicopter or drone hovering in front of a line of trees or a hillside. When the drone starts moving again, however, it is likely to be seen more or less immediately.

Thermal signature depends on the amount of heat given off, and is only relevant if devices that can see infrared radiation are present. The naked eye will only spot heat emissions if they cause a haze or if there are actual flames, and a drone that is emitting flames is probably already in enough trouble. Thermal signature from a small drone with an electric motor is very small, but larger drones, especially those with combustion or jet engines, can produce enough heat to make them very obvious.

Some anti-air weapons home in on thermal emissions, although few drones produce enough to be targeted and most are not worth expending a missile upon. Those that do produce a lot of heat can mask it by design features. For example, placing engines above the tailplane puts a cooler surface between the extremely hot jet exhaust and most observers, which will likely be below the drone. Advanced engines also produce less hot exhaust and dissipate heat more efficiently than earlier models, which also acts to reduce heat signature.

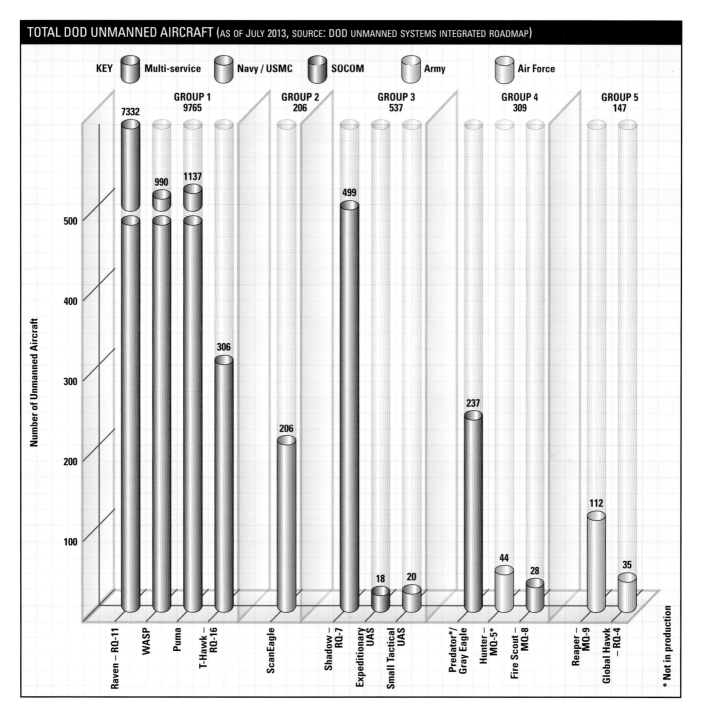

TOTAL DOD UNMANNED AIRCRAFT (AS OF JULY 2013, SOURCE: DOD UNMANNED SYSTEMS INTEGRATED ROADMAP)

KEY — Multi-service — Navy / USMC — SOCOM — Army — Air Force

Number of Unmanned Aircraft

Group	Aircraft	Number
GROUP 1 – 9765	Raven – RQ-11	7332
	WASP	990
	Puma	1137
	T-Hawk – RQ-16	306
GROUP 2 – 206	ScanEagle	206
GROUP 3 – 537	Shadow – RQ-7	499
	Expeditionary UAS	18
	Small Tactical UAS	20
GROUP 4 – 309	Predator*/Gray Eagle	237
	Hunter – MQ-5*	44
	Fire Scout – MQ-8	28
GROUP 5 – 147	Reaper – MQ-9	112
	Global Hawk – RQ-4	35

* Not in production

65

Left: Rotorcraft drones, such as SkyRanger, can be used in very confined spaces, and can enter buildings through a suitable opening. The ability to hover provides a very stable sensor platform. The tradeoff is a high power requirement to keep the craft airborne, which shortens operating duration.

Left: The RQ-2 Pioneer UAV was first deployed aboard warships, but can provide reconnaissance in a range of environments. It can take off from a runway in the manner of a conventional aircraft or be launched from a catapult. Naval versions use a rocket-assisted takeoff.

Audible signature is determined by how much noise a drone makes. Electrically-driven drones are virtually silent, although propellers do make a small amount of noise and can be heard from nearby. Drones using combustion engines are noisier, although advances in technology have produced engines that operate quietly. Some early drones made a noise much like an airborne lawnmower, which tended to advertise their presence somewhat.

Left: The SkyLite UAV can be launched from a canister with its wings folded. Upon clearing the canister, the wings snap out into flight position. It is a short-ranged drone with a duration of about an hour, intended to provide local reconnaissance and situational awareness to ground forces.

Electromagnetic signature depends on the amount of active emissions the drone is making. Radio signals, radar emissions and the like are all detectable at considerable distances to any suitable instrument. This can be countered by using low-powered emissions where possible, and by limiting transmissions to a minimum. Low-probability-of-intercept (LPI) radar equipment has been developed over recent years enabling a drone or other platform to use active radar without giving away its position.

Reducing the level and amount of electromagnetic emissions is not feasible for drones that are acting as radio relays or electronic warfare platforms with active jammers; these overt actions are likely to be detected without difficulty, but for most applications it is possible to keep the drone's electromagnetic signature to a minimum.

Radar cross-section is a measure of how detectable an aircraft or drone is to radar. Small size translates directly to a minimal radar cross-section, but there are other considerations as well. Materials that reflect electromagnetic radiation (e.g.

radar signals) will include most metals; a drone that uses a lot of ceramics, carbon fibre or natural materials like wood will have a smaller radar cross-section than one composed mainly of metal.

Sharp corners and large flat surfaces also reflect radar energy very well, and it is these reflections that are picked up by the detector. Thus stealthy drones attempt to use rounded surfaces and components angled to scatter radar energy rather than reflect it straight back to the detector. This cannot eliminate the chance of detection, but it does reduce the amount of energy being picked up by the detector, thus ensuring that the drone can be detected and tracked only from quite close by.

Stealthy Drones

All of these technologies combine to create drones that are unlikely to be detected and tracked. This has a number of applications. Obviously a drone that is not detected cannot be attacked, but it is also important sometimes that the enemy does not know he is being observed. A stealthy drone can be used to undertake reconnaissance into areas where a detection would lead to an international incident, or where it is simply too risky to operate manned aircraft.

Stealthy design is primarily relevant to larger drones; small hand-launched types are unlikely to be detected by any means other than direct observation, and their small size makes this event quite unlikely. Stealthy design is not always aerodynamic or cost-effective, so often small drones cannot make use of stealth technology, even if it might be beneficial.

Above: The Corax (or Raven) UAV has been referred to as a 'stealth drone'. It incorporates low-observable technologies and appears to be designed for the surveillance/reconnaissance role. However, some observers have suggested that with a different wing fit this UAV might be converted to a high-speed penetrator/strike craft.

Missiles and Bombs

Missiles tend to be fairly large and heavy relative to the average drone, and can be carried only by larger models that have sufficient power to lift them. The AGM-114 Hellfire missile can be deployed from a variety of platforms, including Predator and Reaper drones. It was developed as a precision-guided weapon to be launched from helicopters and similar platforms against mobile or small targets, such as tanks and bunkers.

Most Hellfire variants are laser guided, giving an extremely high degree of precision so long as the designator can be kept on target. As with all semi-active guidance systems, the missile itself makes no radar or similar emissions and the laser is not likely to be detected, reducing the chance that the target will evade in time. The laser can be blocked by smoke or haze, however. A radar-guided version is available, but it is not used with drones;

future versions may be able to make use of a multimode seeker that includes radar and other guidance systems.

The Hellfire missile can carry a variety of warheads depending on its mission. Early versions were envisaged as tank-killers, but the ability to engage other targets is desirable in the modern battlespace. An anti-tank warhead uses a very focused explosive charge to penetrate thick armour, whereas other targets may require a different warhead effect. For anti-personnel and 'soft target' applications, a larger blast radius is desirable. This increases area of effect at the expense of penetration. The Brimstone missile was developed from

Above: The GM-114 Hellfire missile was developed for use aboard combat helicopters as a tank-killing weapon. With laser guidance it is capable of great precision, making it suitable for strikes against a wide range of targets. Hellfire missiles have been used for 'airborne assassinations' of key insurgent personnel.

Hellfire and uses a multimode seeker with both laser and radar guidance. The combination of guidance modes makes the weapon far more resistance to countermeasures – radar can see the target through smoke that blocks the laser, and the laser cannot be jammed by electronic countermeasures. Brimstone missiles have been successfully launched from MQ-9 Reaper drones and have shown that they are capable of hitting even a rapidly moving ground target.

The AGM-176 Griffin missile was developed in response to modern combat conditions, where small and inexpensive, but precise, weapons play an extremely important role. Costs were kept down by using components from other systems, such as the Javelin anti-armour missile and the Sidewinder air-to-air missile. Griffin can be launched from a variety of platforms, including transport aircraft, helicopters and vehicles, as well as UAVs, such as the MQ-9 Reaper.

LAUNCHING A LASER-GUIDED MISSILE FROM AN MQ-9 REAPER

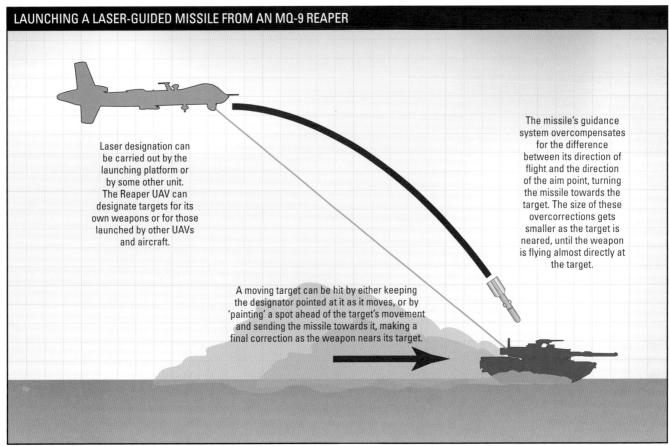

Laser designation can be carried out by the launching platform or by some other unit. The Reaper UAV can designate targets for its own weapons or for those launched by other UAVs and aircraft.

The missile's guidance system overcompensates for the difference between its direction of flight and the direction of the aim point, turning the missile towards the target. The size of these overcorrections gets smaller as the target is neared, until the weapon is flying almost directly at the target.

A moving target can be hit by either keeping the designator pointed at it as it moves, or by 'painting' a spot ahead of the target's movement and sending the missile towards it, making a final correction as the weapon nears its target.

Left: The Brimstone missile started out as a developed version of AGM-114 Hellfire, but has since been almost entirely redesigned. It was originally intended to be compatible with aircraft such as the Harrier, which is no longer in service, but can be deployed from a range of platforms.

Although its warhead is small, the AGM-176 Griffin missile offsets this with precision and, in later models, a multi-effects warhead that can be used against a wider variety of targets. It uses dual-mode guidance combining laser and thermal seeking, and it has a range that is comparable to the larger Hellfire missile system.

Air-to-air missiles have also been mounted on drones, notably the Predator. The proven FIM-92 Stinger was chosen as defensive armament for Predator drones operating in Iraq. Its light weight made it a good choice, and it had been shown to be effective when launched from a variety of platforms, ranging from helicopters to the shoulders of infantrymen.

The Stinger Missile

The Stinger missile is a short-range weapon using infrared homing to attack its target, and as a fire-and-forget missile it was fairly easy to integrate with drone technology. It requires no targeting information from the drone once it is pointed at the target and activated. However, it did not prove effective against hostile fighters. In order to create an effective anti-air platform, it would be necessary to deploy drones that can manoeuvre rapidly to obtain a firing position, which at present is not possible.

Right: AGM-176 Griffin is a small, lightweight weapon capable of precise attacks, making it very suitable for UAV operations. Its small warhead limits the amount of damage it can do, which is actually a good thing when targeting hostiles in proximity to neutrals or friendly forces.

There are still potential uses for drone-launched air-to-air missiles. Helicopters, large enemy drones and slow-moving aircraft, such as, transports could potentially be engaged with Stinger missiles launched from drones, and there is a possibility of using a mobile version of the 'Stinger Ambush' tactic to disrupt enemy air operations.

A Stinger Ambush would normally be staged by positioning personnel armed with shoulder-launched Stinger missiles (or other ground-to-air weapons) on the flight path of aircraft, ideally close to an air base where aircraft taking off or coming in to land can be targeted. This requires considerable effort and risk to get personnel into place, but allows high-value targets to be attacked by surprise or high-performance fighters to be engaged with a good chance of success.

Lightweight anti-aircraft missiles like Stinger have a limited engagement envelope and cannot target high-flying aircraft. They can also be outmanoeuvred by many fast jets, but a fighter that is just climbing off the runway and has little airspeed is much more vulnerable. Thus a drone could be used to make an attack against enemy aircraft in an area that could not be reached on the ground.

Reaper drones can also carry the AIM-9 Sidewinder air-to-air missile. This is another well-proven weapon system, with several versions introduced since they first entered service in 1956. Sidewinder missiles can be guided by infrared or radar, and have been successfully used aboard helicopters. Experience in the latter role shows that air-to-air missiles aboard relatively low-performance platforms are mainly useful against similar targets rather than fast jets. However, the ability of

the Sidewinder missile to lock onto an aircraft from any aspect angle, including head-on, gives these weapons at least some capability against fighters.

Drones can also deliver bombs, although the size of a bomb that can be carried is limited by the lifting capability of the platform. The Reaper drone can carry the GBU-12 Paveway laser-guided bomb, which uses a 227kg (500lb)

Below: The Stinger missile has proven highly effective in a man-portable configuration, and experiments have been undertaken with stinger-armed UAVs. This did not create an unmanned fighter aircraft, but offers the possibility of self-defence or the chance to attack enemy helicopters operating in the same area as the UAV.

warhead. This is small by the standards of combat aircraft, but it is entirely sufficient to destroy most targets that might be targeted by drone strikes.

Laser guidance allows great precision, which permits a smaller warhead to be far more effective than would be the case with unguided 'iron' bombs. A smaller warhead is also desirable in some applications, such as precision strikes against hostiles in proximity to non-combatants or when delivering close support to ground forces. In today's battlespace, smaller is sometimes better.

GBU-38 Bomb
The GBU-38 bomb is also developed from the same Mk82 227kg (500lb) bomb as the GBU-12, but rather than

Right: The AIM-9 Sidewinder was developed as an air-to-air missile, unlike Stinger, which was originally a ground-to-air system. Later model Sidewinders can make an 'over-the-shoulder' launch – the launching platform does not need to be pointed at the target. A Sidewinder-armed UAV could theoretically loiter in an area and ambush enemy aircraft passing through.

using laser guidance, it is fitted with a Joint Direct Attack Munition (JDAM) guidance kit to enable GPS guidance once launched. Although slightly less precise than a laser-guided weapon, GPS guidance does not require a laser to be aimed at the target; the weapon requires no guidance after launch other than the GPS signal that tells it where it is.

Right: The GBU-12 is a 500lb (227kg) bomb using laser and, in later models, GPS guidance. This size of warhead is the smallest in use aboard most strike aircraft, but represents the limit of what can be carried by even the larger UAVs. With good guidance, the warhead is sufficient for most tasks.

Larger bombs can be carried by some drones, including the GBU-16 that is based on the Mk83 454kg (1000lb) bomb that has been in service for numerous years. Although still fairly modest by the standards of strike aircraft, this is a potent weapon capable of destroying most targets. Indeed, there has been a move in recent years towards multiple lighter bombs rather than small numbers of heavy ones.

This is not least because the sort of targets that are engaged in most modern conflicts do not necessitate large warheads. A 454kg (1000lb) bomb is wasteful against a small group of armed

Above: Like GBU-12, the GBU-38 bomb is based on the Mk82 general-purpose 227kg (500lb) bomb. It is GPS-guided and can achieve sufficient precision to be used against targets in proximity to non-combatants. GPS is somewhat less precise than laser guidance, but does not need any further action after the bomb is released.

insurgents or a light, soft-skinned vehicle armed with a machinegun. These targets may also be in proximity to friendly troops or civilians, so, again, a smaller weapon is desirable to avoid collateral damage.

For this reason, and to allow aircraft to carry more munitions, the GBU-39 Small Diameter Bomb was developed. Initially equipped with GPS guidance, the GBU-39 can be fitted with other seekers, including thermal, radar and laser guidance systems or a multimode seeker combining all three. The GBU-39 has a 110kg (250lb) warhead that is more than sufficient for most missions. Its small size

and relatively low weight makes it highly suitable for drone operations.

The GBU-44 Viper Strike bomb is extremely small, with a 1kg (2.2lb) warhead. This limits the amount of damage it can do, which may be a good or bad thing depending upon circumstances. If a greater blast effect is required, heavier weapons are available, but for precise strikes on small targets in proximity to people or property that must not be harmed, the availability of a small weapon of this kind offers new possibilities for drone operators.

The Viper Strike bomb was developed

from anti-armour weapons that, by definition, require a high degree of precision if they are to be effective. It can use a tandem warhead, i.e. one with two explosive charges that fire in quick succession to defeat advanced armour, and has a 'danger close' radius of 50m (164ft). That is to say, Viper Strike can be used with confidence against targets 50m from friendlies without endangering them.

A drone strike using Viper Strike bombs can put a warhead within 1m (3ft 3in) of the target using GPS and laser designation. The weapon is designed to maximize damage to the target while reducing the scattering of fragments and blast damage. It is reported that in urban terrain, Viper Strike bombs will not cause any damage beyond 16m (52ft) from the impact point.

Drone-launched bombs do not have the same range as those dropped from fast-moving aircraft, partly due to drones generally operating at a lower altitude, and partly because the launch platform cannot use its speed to 'toss' the bomb towards the target by launching while climbing rapidly. However, drone-launched munitions are just as precise and effective as their aircraft-dropped counterparts, and may be more quickly available if a drone is kept ready in the area in case it is needed.

Other Weapon Systems

Other weapon systems may eventually be fitted to drones. A laser-guided version of the traditional multiple-launch rocket pod, long a standard weapon for ground attack by helicopters and aircraft, has been available for some time and is being cleared for use aboard an expanding array of aircraft.

Guided rocket pods may thus become a drone weapon system in the near future, whereas unguided rockets are probably not cost effective or suitable for the environment in which most drone operations are carried out. Unguided rockets are by definition rather indiscriminate, being best suited to strafing of a largish area. This is effective in large-scale combat, but a more precise strike is necessary to avoid collateral damage in most drone operations. Besides, drones are only able to carry a limited warload, so it seems desirable to get the most out of the weapons that can be carried. Guidance translates to increased effectiveness per rocket, which may be a critical consideration.

Right: Helicopter-launched rockets have been in service for decades, but until recently they were less than precise. The addition of laser guidance has turned what was originally a strafing weapon into a system capable of striking a specific target even if it is moving.

It is possible that gun pods will be carried by drones in the future. Self-contained gun pods are carried by some aircraft, and the technology could be adapted to drone operations. There have been sightings of drones with pods under the wings that could contain guns. They could also be fuel tanks, or carry instruments rather than weapons, and, if they contain weapons these might equally be missiles. However, the pods look in some cases similar to those used for aircraft gun systems, so it is possible that gun-armed drones are already flying.

It is questionable how much use a gun-armed drone might be, however. Most drones are not very manoeuvrable and they would find it difficult to bring a gun on target with any real effectiveness.

A guided weapon makes much more sense for drone applications – one that can be delivered to a launch point and sent on its way without having to try to keep an unwieldy drone on target. Bombs and missiles offer far more effectiveness for the same investment of effort, although the psychological effect of a drone strafing personnel targets might be significant.

However, for the foreseeable future, it is likely that drone weapon systems will be restricted to small bombs and light missiles, mostly intended for use against precision targets on the ground. The heavy-strike and air-to-air niches are already well filled by combat aircraft; drones will have to show significant advantages before they begin to displace the manned aircraft from these roles.

Combat drones

Combat drones, i.e. those with the capability to deploy weapons as well as sensor packages, are the most common type to make the news and, for the same reasons, the most controversial. Their role in combat zones worldwide has expanded to the point where we are as used to hearing about 'drone strikes' as 'air strikes'. Indeed, to some people the terms have already become interchangeable.

Above: It seems likely that China's Pterodactyl UAV was based upon, or at least influenced by the MQ-1 Predator drones already flying. Whilst the US places strict restrictions on export of certain technologies, China's less stringent regulations may result in these drones becoming available to a wide range of users.

The use of drones for air strikes is not a new concept, but it has taken many years for the necessary technology to be developed. The pitfalls along the way are littered with the wreckage of experimental craft, and, even today, when it is possible for foreign powers to copy existing designs, there are still blind alleys and development issues to overcome.

This was illustrated in 2011, when a Chinese drone bearing a startling resemblance to an RQ-1 Predator crashed for unknown reasons. Designated Pterodactyl, this drone appears to be based on the successful Predator and is intended for a similar role. A large part of developing a new technology is figuring out what can and what cannot be done; Predator had already shown what was possible with this design of drone, which made the Chinese development team's efforts somewhat easier.

However, despite this headstart the process of creating a combat-capable drone was complex and did not always go to plan. Few details of the 2011 crash are available, as the site was rapidly closed off and the wreckage removed,

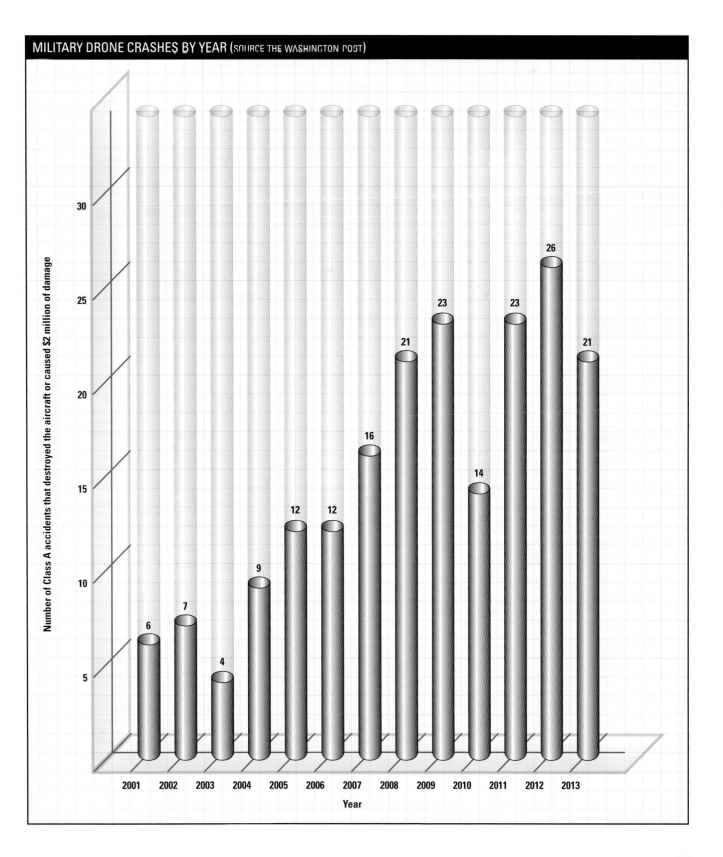

MILITARY DRONE CRASHES BY YEAR (SOURCE THE WASHINGTON POST)

Above: The RQ-2B Pioneer has been in service since the mid-1980s; much longer than many people might imagine. Whilst the US and other long-term UAV operators have gained a significant body of experience using these early drones, newer operators are starting from scratch and are more likely to make expensive mistakes.

but it seems likely that the problem was either a fault with the craft or an error made by the operator.

This is an area where 'borrowed' technology cannot really help. The operation of UAVs from a remote ground station is a new field of endeavour, with rules that have to be learned along the way. The USA and other drone operators have lost their share of drones through operator error or other unexpected consequences arising from procedures that seemed sound when the drones were developed.

One US drone crash occurred when an operator transferred control from one console to another without following the checklist created for such a handoff.

The result was that the drone's control electronics shut down and could not be restarted. Incidents like this can perhaps be anticipated and avoided in an ideal world, but inevitably there will be incidents that were not foreseen. Experience can only be gained through live drone operations, and, in this field, every nation must start more or less from scratch.

The primary benefit that larger, combat-capable drones offer over smaller counterparts is, of course, their ability to deliver an effective attack. However, this comes at the cost of increased price and size. It may be that the benefits of a wider range of missions are outweighed by the ability to afford fewer drones. Numerous quite small drones may offer

more advantages than a few more potent ones. After all, weapons can be delivered by a variety of means, most of which require good reconnaissance that could be provided by small, cheap drones and might be absent in some areas if insufficient numbers are available due to budget constraints.

Thus, the decision to buy large multipurpose drones is not as simple as it may at first seem. They require a greater level of support from both a technical and a logistical standpoint, and many users might not find that the ability to deliver weapons is worth the added cost.

However, for those users that can afford them, combat-capable drones offer many advantages. They are capable of carrying quite large instrument packages and thus gathering a range of data while covering large areas or operating very far from base. Although weapons' use makes the news, long-range drones

Above: The RQ-1 Predator drone flew over 600 missions during the Balkans conflict of the 1990s, and has been highly active in Afghanistan and Iraq. Most of its activities go unreported – 'drone strikes' make headlines but intelligence gathering and reconnaissance go on all the time without anyone noticing – which arguably is the point of UAV operations.

spend a lot of time in the air doing nothing more exciting than monitoring an area. This is important work, but mundane, and only gets reported when something goes wrong or a controversial incident occurs.

The primary advantage of operating this class of drones is the same as any other class – they are a reconnaissance and surveillance asset that can be kept aloft for a long period. However, there are times when the ability to send what is ultimately an expensive but expendable asset into a dangerous area to support ground troops might be highly useful. There are some areas that are too dangerous for helicopters or even strike aircraft, but a drone mission risks no lives. While the cost of the platform is considerable, using a drone means not having to choose between risking pilots on dangerous support missions or leaving ground forces without support.

RQ-1/MQ-1 Predator

Probably the best known of the combat-capable drones, the Predator first flew in 1994 and entered production in 1997. Developed and built by General Atomics

Aeronautical Systems, the Predator was initially designated RQ-1 (R for Reconnaissance and Q designating an unmanned air vehicle), but was eventually redesignated MQ-1 to reflect its multirole capability. It is officially considered to be a medium-altitude, long-endurance (sometimes abbreviated to MALE) unmanned air vehicle (UAV).

The Predator drone uses a 101hp (75kW) Rotax four-stroke internal combustion engine of a similar sort that is used for snowmobiles to power its two-bladed 'pusher' propeller. Developing a suitably quiet engine was one of the challenges encountered in creating military drones, but this was overcome to a sufficient degree that the drone has a low acoustic signature.

The engine is located at the rear of the craft, along with a distinctive downward-sloping tail and a downward-facing rudder. Wings are straight and edged with titanium. Early problems with icing

at high altitudes were overcome by implementing a 'wet wings' configuration, with many small holes continuously 'weeping' ethylene glycol (anti-freeze) onto the wing surfaces.

Although quite large for a drone, the Predator keeps weight down by using advanced composite materials. The main fuselage uses carbon and quartz fibres bonded into Kevlar, with a frame composed of carbon fibre and aluminium.

Internal electronics are powered by an alternator running off the drone's engine, backed up by batteries. An external power feed is used to start the drone's systems, which are then self-powered for as long as the drone has fuel and the engine is running.

The Predator carries most of its sensor package towards the front of the craft. Its communications antennae are located in the bulbous upper frontal fuselage, with other systems, such as synthetic aperture radar, located blow. A gimballed turret

Above: Like any aircraft, the Predator needs regular maintenance and repairs when it is damaged. UAVs often operate from extremely primitive forward bases, and in such conditions damage is inevitable from time to time. Fortunately, repair facilities can be more basic than would be required for large aircraft.

allows cameras to point in any direction independently of the drone's flight path.

For transportation, the Predator UAV can be disassembled and carried aboard a cargo aircraft, such as a C-130 Hercules. The ground control station also has to be transported to the operating area, although it is no longer necessary for the operator to be on the same base that the Predator is flying from. In the first years after deployment, Predator drones were controlled from forward bases, but the advent of improved satellite communications made this unnecessary.

After problems with severe lag between control operation and response from the UAV were solved, a Predator

over Afghanistan could be flown from the mainland USA. However, a launch and recovery team is still required at the beginning and end of the mission and must be deployed along with the drone itself. The usual deployment is four Predators, plus a ground control station and a data distribution terminal.

Normally, the Predator's ground station consists of a mobile trailer containing consoles for the pilot and additional operators. More consoles can be used for data exploitation – i.e. making immediate use of or disseminating information gained from the Predator's sensors – and for mission planning. This enables part of the operating team to be

setting up a new phase of a mission while the operating team is carrying out the current set of orders.

In addition to solving problems with icing, which caused many of the early operational losses, the MQ-1B version incorporated a number of improvements

and some slight aerodynamic changes, including longer wings and a turbocharged engine. Experiments with using Predator drones fitted with missiles were carried out at the end of the 1990s, but it was not until after the 11 September 2001 attacks on the

USA that the concept was implemented operationally.

In November 2002, a Predator drone was used to deliver a Hellfire missile against a motorcade carrying an al-Qaeda leader, Qaed Sinan al-Harethi. The Predator proved highly effective in this

Above: The improved Predator B model made its first operational flights over the Balkans. In 2005, the US Army selected a further developed version, designated Sky Warrior, as part of its extended-range, multi-purpose UAV project. Sky Warrior made its first operational deployments in the late 2000s.

Left: One of the earliest targets for a 'drone assassination' mission was Qaed Sinan al-Harethi, pictured here with fellow al-Qaeda leader Osama bin Laden. A stealthy drone strike permits action against high-value targets that would be inaccessible to ground forces and might be able to evade a strike by noisy conventional aircraft.

Below: Creating strike-capable UAVs was not simply a matter of fitting pylons and installing targeting software. The weapon had to be integrated with the rest of the UAV's systems, and in addition, all-round safe operations had to be considered – landing with live ordnance aboard is always hazardous for any aircraft.

role, combining a quiet approach with a very precise strike using a supersonic missile. In short, the Predator strike was an air-launched assassination that gave the target no chance to react.

Guidance for the Predator's two Hellfire laser-guided missiles was provided by a multi-spectral targeting

system (MTS) that uses infrared and thermal television cameras and a laser designator to provide targeting information. This is combined with data on temperature, wind speed and other environmental conditions to create a detailed firing solution that can be used by the UAV's own weapons or handed off to another platform, such as a strike aircraft or guided artillery weapon.

The Predator's standard electronics package includes a Forward Looking Infrared system, television cameras and synthetic aperture radar. A forward-looking nose camera is primarily used by the pilot to fly the UAV using First-Person View (FPV). All the nose cameras can be used to obtain full-motion video images.

SPECIFICATIONS: RQ-1/MQ-1 PREDATOR

Length: 8.2m (27ft)
Wingspan: 14.8m (48ft 6in)
Height: 2.1m (6ft 9in)
Powerplant: Rotax 914 4-cylinder, 4-stroke, turbocharged engine
Maximum take off weight: 1020kg (2249lb)

Maximum speed: 129km/h (80mph)
Range: 730km (454 miles)
Ceiling: 7620m (25,000ft)
Armament: 2 × AGM-114 Hellfire laser-guided anti-tank missiles or 2 × AIM-92 Stinger short-range anti-aircraft missiles

Above: Like conventional aircraft, UAVs require a post-flight inspection to spot any damage or maintenance issues that might have arisen during the mission. The long duration of UAV flights can be wearing on engines and other components, which are subject to stress for the whole time the UAV is airborne.

In addition, the Predator can carry other electronics packages tailored to the mission at hand. This can include additional cameras and similar sensors or packages tailored to a different mission, such as SIGINT (Signals Intelligence), allowing the Predator to intercept enemy radio traffic and relay it back to base.

Although it was the US Army that first evaluated the Predator drone for service, it is operated by the US Air Force. Predator drones have served in a variety of theatres, including the Balkans, Afghanistan, Iraq and the surrounding territories. In addition to launching strikes and conducting reconnaissance, some Predator drones were used as decoys in the hope of locating the positions of enemy air defence weapons.

In 2005, permission was requested for Predators to be used to assist in the disaster relief operations after Hurricane Katrina hit, but this was not possible because of restrictions in place on unmanned aircraft operations over the USA. In 2006, permission was granted for drones to operate in US airspace under some circumstances, although they are still subject to restrictions. Military Predator drones have been used in search and rescue operations and to help monitor fires, and some have been used as testbeds for the development of new systems.

Although the last Predators have been delivered to the US Air Force, they will remain in service for some time and are also serving with other militaries.

Non-military Predator drones are used by government and law enforcement agencies to monitor borders and other open areas that are difficult to patrol.

MQ-9 Reaper

The MQ-9 Reaper drone began development in 2005 as a variant of the Predator B model, at the time designated Sky Warrior. The US Army was interested in an extended-range multirole UAV, while the US Homeland Security/Customs & Border Protection agency also wanted a UAV for patrol work. Armament was not a consideration for the latter. During the same period, contracts were awarded for system design and development of the MQ-9 Reaper Hunter-Killer drone.

Early Sky Warrior drones were deployed in 2008, operating in Iraq, and, in the same year the US Air Force began operating the MQ-9 Reaper in

Afghanistan. The US Air Force had formed its first Reaper squadron in 2007, and by the end of that year was operating early-model Reapers in Iraq.

The MQ-9 Reaper is visually similar to the Predator. The easiest way to tell them apart is the tail section – the twin tailplanes point diagonally down on the Predator and up on the Reaper. Propulsion uses a similar rear-mounted pusher propeller, but this is driven by a vastly more powerful 950hp (708kW) turboprop engine. This allows an enormously increased payload, increasing the UAV's utility as a weapons' platform, and increases performance in flight.

The Reaper UAV uses many systems developed for the Predator. There is nothing outmoded or obsolescent about this; several electronic systems have matured and been upgraded during their service with the Predator and remain at the cutting edge of military electronics.

The first Reaper UAVs used an enlarged version of the Predator airframe that served to prove some of the concepts inherent in its design. A variant

Below: The Predator B first flew in 2001 and is in service with the US Air Force and the Royal Air Force as MQ-9 Reaper. It is twice as fast as the original Predator UAV and, more importantly perhaps, carries five times the payload to twice the altitude.

powered by a turbofan (jet engine) rather than a turboprop engine was trialled, along with an enlarged airframe that retained turboprop propulsion. This latter was designated Altair and went into service with NASA.

The Altair UAV diverged from the Reaper in several ways. It had no need for weapons' capability and, as such, it was never given any, but it did receive an upgraded avionics package. This was in part to fit its role as a testbed aircraft for advanced sensor equipment and in part to comply with Federal Aviation Authority (FAA) rules on flying unmanned air vehicles in US airspace.

Below: The Reaper (Predator B) is most easily distinguished from a first-generation Predator by its tail. The original Predator UAV had two tail projections angled downward; Reaper has two angled upward and a third directly downward. Both are propeller driven; a jet engine and two upward-pointing tail projections indicates a Predator C (Avenger).

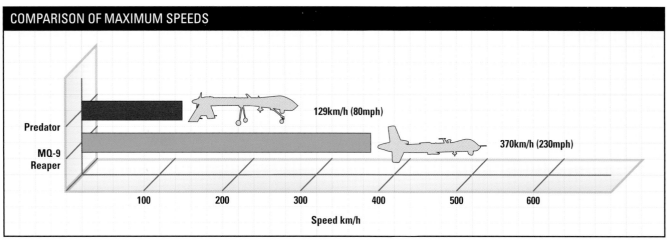

COMPARISON OF MAXIMUM SPEEDS

Predator — 129km/h (80mph)

MQ-9 Reaper — 370km/h (230mph)

100 200 300 400 500 600

Speed km/h

Above: NASA's Altair UAV is a modified Predator A used to develop and demonstrate technology to be deployed aboard high-altitude unmanned science platforms. The project had a parallel goal of facilitating UAV use in civilian airspace. Although some progress was made, this goal was not achieved.

NASA uses the Altair drone to conduct Earth Sciences experiments. Its thermal cameras proved useful in mapping the Esperanza Fire in California in 2006. This use of a former military drone to assist in disaster response was a first; having demonstrated the capabilities of drone aircraft in providing such assistance it seems likely that the practice may become commonplace in the future.

In the meantime, the MQ-9 Reaper went into service with the US Air force and has since served in Afghanistan, Iraq and other theatres. It has successfully used both bombs and missiles in strikes on insurgent forces, using precision guidance to attack vehicles. This service was not without incident. The Reaper and Predator drone fleets have been described as being highly accident-prone, which is not unfair given their record.

One reason for the number of incidents is possibly reliance on technology, which is never perfect. Control lag was largely eliminated during the early phases of Predator operations, but that does not rule out the possibility of the occasional glitch or signal interruption. This may have been the reason for the 2009 loss of a Reaper over Afghanistan.

In this incident, control over the drone was lost, creating a situation where an armed aircraft was flying around with unknown intentions. The only recourse was to shoot it down, so a fighter was sent to disable the drone with a missile. Ironically perhaps, control was restored just after the Reaper's engine was disabled, requiring a deliberate crash to prevent the drone from being a hazard to anyone in the vicinity when it came down.

Other Reapers have crashed due to mechanical failure or for unknown reasons. What is surprising perhaps is that there have not been more pilot-error incidents; flying a UAV via a camera in its nose has been described as being like piloting an aircraft while looking through a straw. The lack of sensory cues, such as sound and inner-ear sense of orientation – not to mention g-forces – can also rob the pilot of important sensory data; he cannot feel what the UAV is doing and while this sort of seat-of-the-pants flying is largely supplanted by instrumentation, the sensory feedback is useful.

The MQ-9 Reaper was the first UAV to equip an air force squadron that flew only pilotless vehicles, but is not intended to replace manned aircraft. The two have

the potential to complement one another, with drones clearing away air defences ahead of a strike and collecting target data for strike aircraft that can carry far more ordnance.

The ability for a UAV to loiter quietly in a target area waiting for a target to present itself is a capability that manned aircraft do not have to anything like the same degree; a manned aircraft can hang around for a few hours – but will almost certainly be detected – but a UAV can be kept over the target area for much longer, with relays of operators taking control of the same aircraft.

The Reaper's warload is much greater than that of its predecessor. It can carry up to four Hellfire missiles and two 227kg (500lb) bombs of various types (usually GBU-12 laser-guided bombs or GBU-38 GPS guided munitions), or similar ordnance. Other weapons include electronic warfare packages; the MQ-9 Reaper has demonstrated in exercises that it can act as an electronic warfare platform in support of manned aircraft and be integrated with their operations.

In addition to a similar reconnaissance role as the preceding Predator drone and an enhanced strike capability, the

Below: The Reaper UAV can carry four Hellfire missiles plus two 227kg (500lb) bombs. These can be GBU-12 laser-guided bombs or GBU-38 Joint Direct Attack Munitions (GPS guided) bombs. Despite all this firepower, it is the sensor turret under the UAV's nose that provides its most useful capability.

Right: Befitting its name, RQ-4 Global Hawk has set records, including the first flight from the USA to Australia by an unmanned aircraft and an endurance record for maintaining a very high altitude (up to 18,288m (60,000ft) for over 33 hours. It has proven to be effective in both military and non-military applications.

Reaper can also serve in other roles. A maritime variant, designated Mariner, has been developed. It had an arrester hook for deck landings, folding wings to save space aboard ship and an increased endurance that may have been as much as 50 hours of flight time.

The Mariner variant lost out to the RQ-4N Global Hawk, but another maritime version, designated Guardian, has entered service (in very small numbers) with the US Coast Guard and Customs and Border Protection se.rvice. These fly primarily counter-narcotics patrols, assisting Coast Guard units and law enforcement personnel to spot and track suspect watercraft.

The MQ-9B Guardian variant has upgraded sensors including an inverse synthetic aperture radar system that is also capable of warning the drone's operators of hazardous weather conditions. The Guardian's airframe and avionics have also been upgraded.

Reaper drones are in service with several countries, including the UK, France and Italy, and have seen service against Islamic State insurgents. British

SPECIFICATIONS: MQ-9 REAPER

Length: 11m (36ft 1in)
Wingspan: 20.1m (65ft 9in)
Height: 11m (36ft 1in)
Powerplant: 1 x Honeywell TPE331-10GD turboprop engine
Maximum takeoff weight: 4760kg (10,500lb)
Maximum speed: 370km/h (230mph)
Range: 1852km (1150miles)
Ceiling: 15,240m (50,000ft)
Armament: Combination of AGM-114 Hellfire missiles, GBU-12 Paveway II and GBU-38 Joint Direct Attack Munitions

Reapers can use the Brimstone missile, which has proven very effective in tests.

Gray Eagle

In 1989, the US Army and Marine Corps jointly began seeking a UAV to provide battlefield reconnaissance and artillery-spotting capabilities. The chosen system was designated RQ-5 Hunter, and was bought in very small numbers before the project was terminated. The Hunter drone entered service in 1996 and was employed in the Balkans and Iraq. The Hunter UAV used a 'heavy fuel' engine, in keeping with the US Army's policy of having only a single type of fuel in the battle area. This simplifies logistics, but does rule out some UAV systems.

The RQ-5 Hunter was a success despite its limited numbers. In 2007, it was used to deliver a laser-guided bomb, the first use of an armed drone by the US Army. Larger and improved versions were purchased over time, and the Hunter was used to demonstrate the capabilities of what would become the Viper Strike precision weapon.

By 2002, the US Army was seeking a replacement. Among the candidates was the Hunter II and a UAV at the time designated Warrior or Sky Warrior. The latter was developed from the RQ/MQ-1 Predator. The US Army had evaluated the RQ-1 Predator UAV with a view to adopting it for service, but the Predator eventually went to the US Air Force.

The Warrior UAV was eventually adopted under the designation RQ-1C Gray Eagle as part of the US Army's Aviation Modernization Plan. It is an obvious derivative of the Predator, with the same airframe design and downward-angled tailplanes, but it uses a heavy fuel engine that can run on diesel or jet fuel.

Below: The RQ-5 Hunter UAV first served in the Balkans in the late 1990s, and later in Iraq and elsewhere. It was to be replaced by the RQ-7 Shadow, but was retained for its greater payload capacity and endurance. Hunters have also served in a national security role.

The Gray Eagle UAV was designed for reliability and fault tolerance, with triply-redundant avionics. This means that exactly the same component must fail or be damaged three times before the aircraft cannot be controlled. That much damage (or sufficient bad luck to cause such a failure) would probably destroy the entire drone.

Gray Eagle incorporates lessons learned with earlier drones, and has a de-icing system built into its wings. Since the vehicle can operate at 8840m (29,000ft), this is necessary to prevent icing and subsequent loss of the craft. An automated takeoff and landing system is also incorporated, simplifying the pilot's task and reducing the possibility of an error at a critical time.

Overall, the reliability of Gray Eagle has been good. The automatic landing system has successfully brought drones home despite strong crosswinds. However, reliability issues have been introduced by retrofitting of new systems. These were mainly caused by software problems, which were eliminated by better integration of the new devices – notably sensor equipment – with the UAV's existing systems.

The Gray Eagle drone is capable of a wide range of missions, including general reconnaissance, damage assessment, convoy escort and communications relay. The recent upsurge in the amount of improvised explosive device (IED) attacks in recent years has led to a search for countermeasures, and the use of persistent reconnaissance drones is one such. Gray Eagle can monitor an area for long periods and detect suspicious activity even at night or through bad

Left: The MQ-1C Gray Eagle UAV is an evolution of the original Predator, created to fulfil a US Army requirement for a long-range multi-role UAV. Its increased wingspan and use of a heavy fuel (diesel) engine were prompted by a desire for improved high-altitude performance.

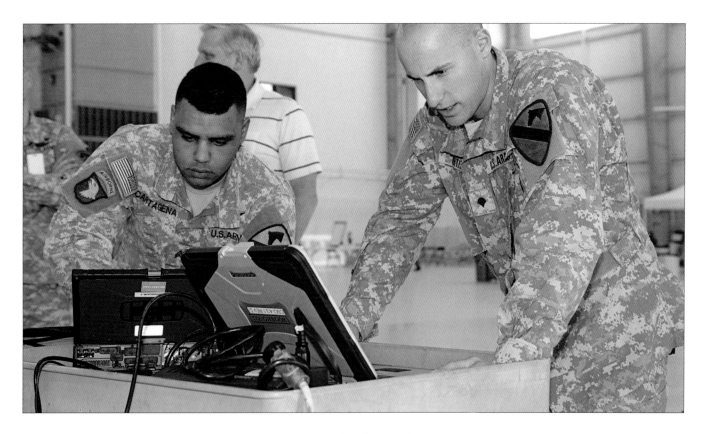

Above: Data from a Gray Eagle drone is carefully studied during testing. The army has slightly different requirements for its UAV fleet to other services, including mine and IED detection and other missions requiring the integration of specialist sensors into a drone's equipment package.

visibility using thermal or radar imaging.

While deterrence is always useful, the ability of a quiet UAV to monitor an area enables more robust countermeasures to be taken. The whole point of an IED attack is that the device must be planted covertly, so insurgents will not act if they think they are being watched. Rules of engagement being what they are, Western forces are required to be very sure that potential targets are hostile. Observing someone in the act of planting a bomb is more than enough proof.

Once the target is confirmed, it must be engaged. Too much of a delay will allow the hostiles to move off or move among civilians where an attack becomes problematic. This is one advantage of using armed drones for observation – a platform in position to observe is also in position to attack.

Drones have been used to attack groups of insurgents engaged in planting an IED on several occasions, and Gray Eagle has this capability using four Hellfire missiles or four Viper Strike munitions. It can also carry Stinger missiles, although these are unlikely to be used against insurgent targets, as such groups tend to lack air power.

Alternatively, the Gray Eagle UAV can carry two Hellfire missiles and an electronic warfare pod or signals intelligence intercept unit. It can remain on station for up to 35 hours in this configuration, far more than is needed to support a single air strike. This capability enables the Gray Eagle to offer coverage for long periods, which might allow several strikes. Another option is the jamming of enemy signals to prevent detonation of radio-detonated IEDs.

The US Army is experimenting with Manned-Unmanned Teaming (MUM-T) operations, whereby a UAV is teamed up with a manned helicopter or other aircraft to provide an additional sensor platform. This allows a suitably equipped aircraft, such as an AH-64E Apache helicopter, to use the Gray Eagle's sensors to search for targets and even designate them for the manned aircraft. This makes it possible to attack from a greater range without exposing the helicopter to return fire.

Gray Eagle is deployed by the US Army in companies that operate 12 aircraft, plus associated control and support systems, and is also used by some special operations units. Hopes have been expressed that every US Army division will eventually have a Gray Eagle company, as the UAVs have proven extremely useful since their introduction.

Avenger & Sea Avenger

Developed from the Predator and Predator B (Reaper) UAVs, Avenger is also designated Predator C, and uses many concepts proven in earlier models from the same family. However, although fuselage shape is generally similar, the Avenger drone has developed considerably from its origins.

One of the most significant changes is the move to jet propulsion. Avenger is powered by a turbofan that is also used in several light passenger aircraft. In addition to increasing payload, this gives a very significant speed increase over earlier designs. By comparison, Avenger has a maximum speed of 740km/h (460mph) as opposed to 480km/h (298mph) and 210km/h (130mph) respectively for the propeller-driven Reaper and Predator UAVs. Avenger has a maximum takeoff weight almost eight times that of a Predator, although, of course, not all of this can be lifted as payload, as the UAV and its engine are significantly heavier.

Nevertheless, the Avenger can carry more at a higher speed than its predecessors. The latter is useful under circumstances where the UAV cannot be operated from local bases close to its operational area. The ability to remain in the air for 30 or 40 hours is less impressive when a drone must fly thousands of kilometres to carry out its mission, then return.

Below: Among the emerging doctrines is the concept of manned-unmanned teaming, where UAVs are integrated with manned aircraft. A UAV is less likely to be detected than a helicopter, and can locate or designate targets. This permits the helicopter to make a 'standoff' attack without exposing the personnel aboard to undue risk.

With slower drones, the only way to fully exploit long loiter times was to forward-deploy the drones to minimize the time transiting to and from their base. Avenger, on the other hand, can cover the same distances much quicker and has a similar endurance, leaving more flight time for the actual mission. Friendly land bases are far more likely to be available within a reasonable operating area from the intended target, or the UAVs can operate from aircraft carriers located off the coast.

Avenger is configured in a similar manner to many 'stealth' aircraft, with the smooth curves and no vertical rudder. Instead, the tail section takes the form of a 'V', with the angled control surfaces acting as 'ruddervators' – a combination of rudder and elevator. This makes manoeuvring more complex, but that does not present a problem for modern avionics.

A human pilot directly controlling a set of ruddervators might struggle to keep the craft under control, but the Avenger's systems translate the pilot's instructions to the aircraft into control impulses, handling the details while the pilot deals with the wider issues of where he wants the drone to go and how it is to manoeuvre.

The tail section is designed to reduce radar return, while the jet exhaust is 'S' shaped to conceal the hot exhaust gases and reduce thermal signature. This cannot be entirely eliminated, of course, but when a stealthy jet-powered aircraft is using relatively little thrust its thermal signature is relatively low. These features create what the media likes to call a 'stealth drone' well suited to high-risk operations in hostile airspace.

The Avenger UAV carries a retractable turret containing thermal and visual cameras, plus a multi-mode radar system that can function as a Synthetic Aperture Radar system or conduct ground moving target indication. In addition, it can carry equipment packages tailored to communications relay missions or electronic surveillance/signals intelligence operations.

Below: Originally designated Predator C, the Avenger first flew in 2009. Advances on the earlier Predator and Reaper drones include jet propulsion and a stealthy design, with weapons carried in an internal bay. Avenger can also carry a greater variety of weapons, some of which are more potent than anything a Reaper can carry.

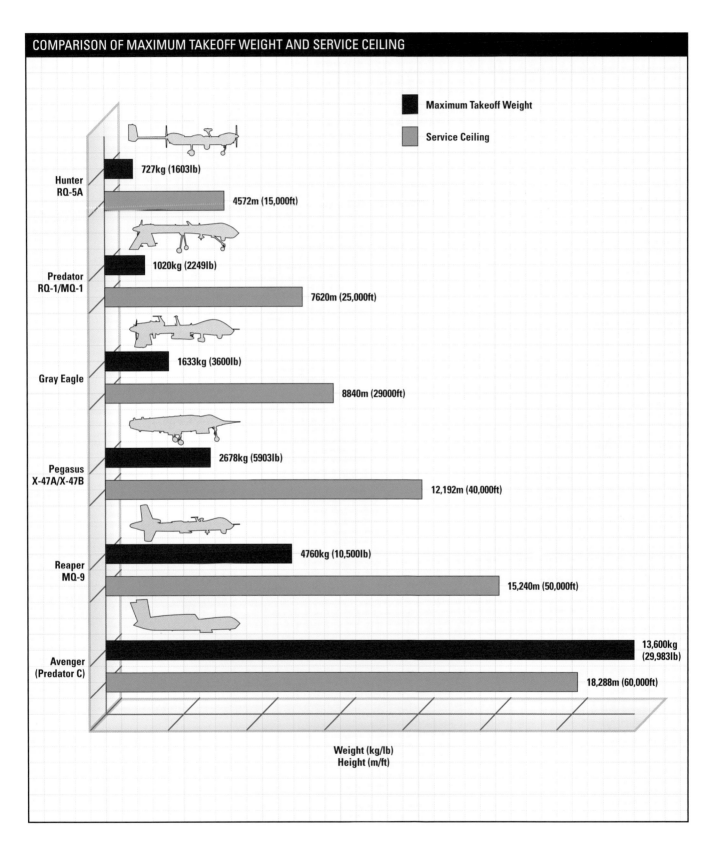

COMPARISON OF MAXIMUM TAKEOFF WEIGHT AND SERVICE CEILING

■ Maximum Takeoff Weight

■ Service Ceiling

Hunter RQ-5A
727kg (1603lb)
4572m (15,000ft)

Predator RQ-1/MQ-1
1020kg (2249lb)
7620m (25,000ft)

Gray Eagle
1633kg (3600lb)
8840m (29000ft)

Pegasus X-47A/X-47B
2678kg (5903lb)
12,192m (40,000ft)

Reaper MQ-9
4760kg (10,500lb)
15,240m (50,000ft)

Avenger (Predator C)
13,600kg (29,983lb)
18,288m (60,000ft)

Weight (kg/lb)
Height (m/ft)

Length: 12.5m (41ft)
Wingspan: 20.12m (66ft)
Powerplant: 1 x Pratt and Whitney PW307
 engine
Maximum takeoff weight: 13,600kg (29,983lb)
Maximum speed: 745km/h (463mph)
Ceiling: 18,288m (60,000ft)
Armament: Can consist of any of the following:
 AGM-114 Hellfire anti-tank missile, GBU-24
 Paveway III guided bomb, GBU-31 JDAM
 guided bomb, GBU-38 small diameter bomb

Avenger can carry more and heavier weapons than its predecessors. Total load is 1588kg (2500lb) of weaponry in the internal bay, plus additional munitions on six wing pylons. Like the Reaper, Avenger can carry Hellfire missiles and bombs ranging from the 113kg (250lb) GBU-39 Small Diameter Bomb to GBU-32 454kg (1000lb) bombs. In addition, Avenger can deploy the 907kg (2000lb) GBU-31 JDAM guided bomb, enabling it to attack larger and better-protected targets than other drones that are in the same family.

For increased stealth, the Avenger UAV can fly with a reduced warload carried entirely in the internal weapons bay. Missiles and bombs carried on angular external pylons produce a much greater radar return than the smoothly curved shape of the Avenger's fuselage, creating a tradeoff between detectability and attack capability. Against an insurgent force, which is not likely to have much in the way of sophisticated radar equipment, this is unlikely to make much difference, but against the military of an

Right: At some point it becomes desirable to stop concealing the capabilities of a weapon system and to start flaunting them instead. Displaying this Avenger UAV, surrounded by some of the very potent weapons it can carry, shows the public what their tax dollars are buying and offers a broad hint to potential aggressors.

advanced major nation, the ability to fly with an absolutely minimal radar cross-section might be vital to mission success.

Avenger was designed to be compatible with ground control stations created for the Predator and Reaper UAVs, but recently a new advanced cockpit control station was introduced, with improved controls and data management systems. In theory, such systems are intended to make the task of the operators easier, but, more commonly, the level of difficulty remains much the same – the price for increasing the complexity of tasks that can be accomplished. Thus the new control system will make routine operations with the Avenger easier, but may also make it possible to carry out new and even more complex missions.

Avenger UAVs are just making their appearance. Indeed, when the US Air Force perceived a need for a stealthy drone to operate in the Afghanistan/Pakistan/Iran region, the Avenger was one of the prototypes that was deployed. This decision proved to be far-sighted; just a few weeks later, an RQ-170 Sentinel drone was lost over Iran. US authorities stated that they were still committed to observing Iran with covert UAV flights, and the increased capabilities of the Avenger may be ideal for this mission.

A variant of the Avenger drone, named Sea Avenger, was developed for the US Navy as part of a programme to create

Below: A Sea Avenger UAV makes a flyby for visiting 'brass', in this case Chief of Naval Operations Admiral Greenert. An increasing proportion of the services' budgets are being channelled into UAVs, making this an area of great interest for senior officers.

an unmanned strike and reconnaissance capability. Sea Avenger features an arrester hook for deck landings and folding wings to save space when stowed away. Its long range and high speed are necessary for nautical operations; virtually every air mission has to cross a lot of empty sea to reach its target area, and many have to fly far inland afterwards.

Sea Avenger may be able to carry a buddy-refuelling system – this enables one drone to refuel another, then fly to base. This will make extremely long-range or extended-duration missions possible, although air-to-air refuelling might be a challenge for pilots and their supporting electronics.

The navy's Unmanned Carrier-Launched Airborne Surveillance and Strike (UCLASS) project is still in its infancy. Indeed, requirements have changed more than once as this new concept is explored. There are benefits to be gained from simply deploying a few UAVs aboard an aircraft carrier in place of an aircraft or two, but, in the long term, a doctrine must be evolved for the use of UAVs in order to use them to their full potential.

It is possible that the world's navies may make use of a similar idea to the US Army's Manned-Unmanned Teaming concept, using a mix of drones and conventional aircraft to complement one another. Other applications include the use of UAVs to support helicopters during amphibious operations. In this case, the UAVs may replace aircraft such as the AV-8B used by the US Marine Corps. More drones can be carried for the same amount of space aboard ship, which may prove advantageous.

Right: The AV-8B has provided air support for US Marine Corps operations ashore for many years. The F-35B was the expected replacement, but it may be that UAVs take on some or all of this role, enabling more craft to be carried aboard a warship of a given size.

It is also possible that some navies that cannot operate full-sized fixed-wing aircraft may field small 'drone carriers'. As the adage goes: 'Any air power is better than no air power', and a small 'drone carrier' would be able to conduct reconnaissance, anti-piracy and some strike operations for a fraction of the cost of a full-sized vessel. From a force-protection point of view, the presence of a small drone force aboard, say, a converted replenishment ship would provide early warning and missile-guidance capabilities.

It is also possible that UAVs could be operated from other vessels, such as tankers and container ships. At times, army and marine detachments have been placed aboard tankers to protect them, such as during the Iran–Iraq War of the 1980s. It is even possible that we may see a resurgence of the catapult-launched fighters used to protect convoys in World War II. Such a capability would be primarily useful for protection against piracy or as an expedient to get some air capability into a region, but it could also be used to create a highly capable search and rescue platform on a fairly small and inexpensive hull.

The capabilities of the sea-borne UAV are only just beginning to be appreciated. It is likely to be the Sea Avenger that shows what can be done, although there are other projects emerging that might offer similar capabilities.

X-47A/X-47B Pegasus

The X-47 Pegasus drone is a futuristic-looking craft shaped much like a manta ray. The original proof-of-concept vehicle was designated X-47A Pegasus in 2001. Its first flight was in 2003. The drone was developed as a private venture with the intention of meeting near-future needs for a naval or air force combat UAV

As with many joint projects, not just those involving the US Air Force and US Navy, the two partners decided that their needs diverged and the joint combat drone project broke up. However, the X-47A was sufficiently promising that an enlarged version was developed, designated X-47B Pegasus, to meet the needs of the US Navy. The X-47B began test flights in 2011 and made its first deck landings later in the year.

Catapult launches and the rapid deceleration of deck landings require a solidly built craft with a strong undercarriage and a frame capable of dealing with repeated heavy stresses. The designers also had to be aware of other considerations when operating at sea – salty air is corrosive and the constant

Below: The X-47B Pegasus UAV first flew in 2011, as part of the US Navy's Unmanned Carrier-Launched Airborne Surveillance and Strike (UCLASS) programme. Pegasus has proven that a UAV is capable of carrier operations, though the UAV finally selected for this role might well be a completely different design.

Above: The X-47 Pegasus uses a diamond-shaped 'flying wing' concept, with different wing designs between the A and B models. Yaw control is performed using elevons on the wing, eliminating the need for a vertical stabilizer (fin) and thus greatly reducing the UAV's radar-cross-section.

activity aboard an aircraft carrier can result in minor dings that can damage a less than robust craft. Thus Pegasus had to be built tough as well as stealthy.

The stealthy shape of the Pegasus UAV presented the designers with a number of technical challenges. An aircraft of any type moves in three dimensions. Pitch (upward and downward movement of the nose and tail) can be controlled by elevators on the wings and tail – if the craft has one. Roll (one wing moving up and the other down) is similarly controlled. Yaw (sideways movement of the nose in one direction

and the tail in the other) is traditionally controlled by a vertical stabilizer or fin – if the craft has one. An angled tail section can provide stability and control over yaw as well as pitch and roll, but this craft design had nothing of the sort.

This was not the first aircraft designed as a 'flying wing', of course, so there was a significant body of knowledge to draw upon. Pegasus solves the yaw

problem with elevons on the upper and lower surfaces of the wing, which make constant small adjustments as directed by the craft's avionics. The elevons are augmented by four small flaps on the top and bottom of the wings. Without constant rapid and complex adjustment the craft would go out of control; an unassisted human pilot could not keep a craft such as this in the air.

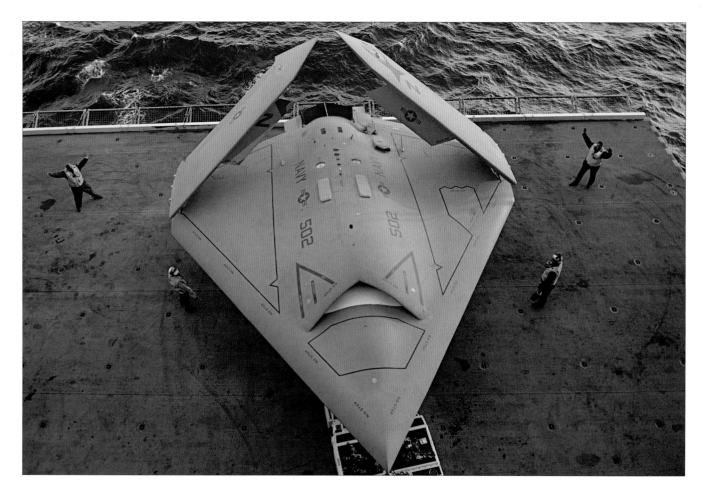

Above: Carrier operations impose a new set of requirements, including the ability to resist salt-water corrosion and hard landings caused by the use of an arrester hook. Folding wings reduce the amount of space the UAV takes up on board, but must be robust enough to withstand flight operations.

Technology has permitted the deletion of fins and subsequent reduction in radar signature, but radar reflections from the engines are a common problem with combat aircraft. The engine position, on the dorsal surface of the drone, reduces radar reflections from its frontal surfaces while still allowing sufficient air to be taken in through the frontal duct.

The engine itself is a high-bypass turbofan of a sort used on light aircraft, and is a proven technology. It provides the UAV with sufficient thrust to be capable of a modestly high speed. Exact specifications have not been released,

but are rated as 'somewhere in the high subsonic range'.

Pegasus is a combat drone, not a reconnaissance platform, and does not carry a complex electronics package beyond what is required for flight operations. Its warload is modest – two bays can each carry a 227kg (500lb) bomb or an equivalently sized weapon – but given its small size and stealthy nature this may well prove to be a highly effective warload.

Besides, this is the beginning of a new technology – pure combat drones have not previously been fielded – and at

present the X-47B is a 'demonstrator', showing what can be done and investigating what might be possible. Future versions of the craft, or others developed from knowledge gained with it, will certainly acquire greater capabilities as the concept and its associated technologies mature.

RQ-5A Hunter

The RQ-5A UAV began development as a result of a joint programme between the US Army and Marine Corps. Initiated in 1989, the project resulted in a contract for small numbers of Hunter drones from 1993 onwards, with the UAV entering US service in 1996. It has since also been acquired by Belgium and France.

Hunter UAVs served in the Balkans in 1999, as part of Operation Allied Force

DEVELOPMENT OF THE FLYING WING

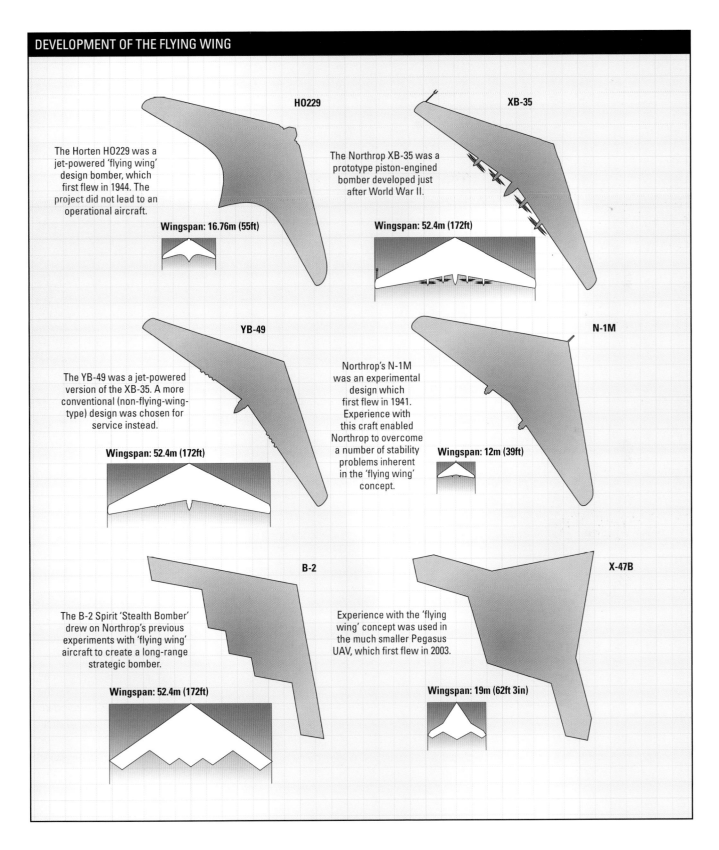

HO229

The Horten HO229 was a jet-powered 'flying wing' design bomber, which first flew in 1944. The project did not lead to an operational aircraft.

Wingspan: 16.76m (55ft)

XB-35

The Northrop XB-35 was a prototype piston-engined bomber developed just after World War II.

Wingspan: 52.4m (172ft)

YB-49

The YB-49 was a jet-powered version of the XB-35. A more conventional (non-flying-wing-type) design was chosen for service instead.

Wingspan: 52.4m (172ft)

N-1M

Northrop's N-1M was an experimental design which first flew in 1941. Experience with this craft enabled Northrop to overcome a number of stability problems inherent in the 'flying wing' concept.

Wingspan: 12m (39ft)

B-2

The B-2 Spirit 'Stealth Bomber' drew on Northrop's previous experiments with 'flying wing' aircraft to create a long-range strategic bomber.

Wingspan: 52.4m (172ft)

X-47B

Experience with the 'flying wing' concept was used in the much smaller Pegasus UAV, which first flew in 2003.

Wingspan: 19m (62ft 3in)

and in Iraq from 2003 onwards. Belgian Hunters were deployed to the Congo in 2006 as part of European military operations there. The US Homeland Security agency also uses Hunter UAVs for border patrol in Arizona.

Although primarily a reconnaissance platform, the upgraded RQ-5B Hunter B can be armed with Viper Strike precision munitions. In 2007, it became the first US Army UAV to deliver an attack on a live target. This version of the Hunter drone first flew in 2005, and offered significantly greater endurance than the original model. It can remain aloft for up to 21 hours as opposed to 12 hours for the RQ-5A.

Also first flying in 2005, an enlarged version named Extended Hunter (or E-Hunter), has a different tail section and a longer wing carrying more fuel. This version can remain in the air for 30 hours and can reach greater heights than B-model Hunters.

The MQ-5B is driven by two diesel engines, located at opposite ends of the fuselage. These power one 'tractor' propeller and one in a 'pusher' configuration. This rests between the twin tail sections. When launching in a small space, such as aboard a ship, Hunter can be boosted by a RATO (Rocket Assisted Takeoff) unit. It can land on a short runway or strip of grass. Belgian

SPECIFICATIONS: RQ-5A HUNTER

Length: 6.8m (22ft 3in)
Wingspan: 8.8m (29ft)
Height: 1.7m (5ft 58in)
Powerplant: 2 x Moto-Guzzi 4 stroke, 2-cylinder pusher-puller gasoline engines
Maximum takeoff weight: 727kg (1,603lb)
Maximum speed: 204 km/h (127mph)
Range: 260km (162 miles)
Ceiling: 4572m (15,000ft)
Armament: 1 x GBU-44/B 'Viper Strike' munition

Below: The RQ-5A Hunter UAV began as a joint requirement for US Army and Marine Corps service. It has since been used by security agencies for border patrol and by overseas buyers in a military capacity. Hunter was the first US Army UAV to attack a 'live' target.

Hunters are fitted with an automated takeoff and landing unit.

The Hunter UAV carries a multimission sensor package with television and forward-looking infrared cameras. Additional payloads can include communications equipment, a laser designator and electronic warfare packages. In addition to its sensors and electronics, it has two hardpoints that can carry a pair of Viper Strike guided bombs.

The RQ-5 Hunter is operated by a two-person crew from a ground station that can fully control one UAV or handle two in relay. It is divided into a pilot bay for flight operations and an observer control bay for payload operations. A third

Above: The MQ-5B uses two heavy-fuel engines and has a greater fuel capacity than the 5A version. It can carry a GNU-44B Viper Strike laser-guided munition under each wing. Viper Strike is an unpowered bomb using GPS and laser guidance. It has a small warhead, but is capable of very fine precision.

bay is used for navigational control, with an optional fourth dealing with intelligence operations and data processing.

The Hunter proved successful in the field, although an accident caused the Belgian B-Hunters to be grounded. It is to be replaced by the more capable Gray Eagle UAV, but this process will take time. Since the existing Hunter fleet remains operable, it is possible that it will remain in service after its official replacement date, perhaps with reserve or second-

echelon units, or for familiarization with drone operations during training.

The Hunter has been used as a payload demonstration vehicle for equipment that might be destined for other UAVs, so another possibility is that after retirement from combat duty the RQ-5 may continue to serve as a developmental testbed. This role is not uncommon for obsolescent aircraft, so unmanned air vehicles may well go the same route.

Extremely long-endurance reconnaissance drones

Strategic reconnaissance can yield extremely useful information about what is going on within the borders of a foreign country or in a very remote area. It can give advance warning of military preparations undertaken in secret, or reveal projects that the host government would prefer to remain secret. These might range from testing a new weapon system in an area that foreigners can be barred from, to the creation of illegal production facilities, such as chemical weapons installations or nuclear weapon projects.

Above: The U-2 strategic reconnaissance aircraft has been in service since 1957, and has received numerous upgrades since. Retirement has been planned for some time, with the Global Hawk UAV as a possible replacement, but thus far the U-2 has shown itself to be too valuable to phase out.

Such data is, of course, virtually impossible to obtain on the ground, no matter what spy movies would have us believe. Satellites offer many possibilities, but these have not replaced aircraft as the primary means of observing distant areas. Strategic reconnaissance platforms include the SR-71 Blackbird and the rather more prosaic U-2 spy plane.

The U-2 has been in service, with many upgrades, for over 50 years. It is a very remarkable aircraft, although tricky to fly and difficult to land. Its ability to fly very high makes it difficult – but by no means impossible – to intercept. An unmanned high-altitude reconnaissance platform could be smaller and thus harder to detect, and would not expose pilots to the risk of being downed while engaged in what might be considered hostile airborne espionage.

Thus the requirement exists for a High-Altitude, Long-Endurance (HALE) UAV to undertake strategic reconnaissance and other missions carried out by aircraft like the U-2. These missions included high-altitude research and testing of equipment or systems to be used in future aircraft or space technologies. One obstacle to the development of such a UAV system was the existence of aircraft that could already do the job. While long in the tooth, these aircraft had the advantage that their development costs were in the distant past; the price tag on an upgrade or new piece of equipment was far less than that associated with developing what was essentially a whole new kind of aircraft.

Right: The SR-71 Blackbird was designed as a replacement for the U-2, but, despite an incredible performance in terms of height and speed, was outlived by its predecessor. The SR-71 was retired (for the second and final time) in 1998, but its role as an extremely high-altitude, long-range reconnaissance platform remains viable.

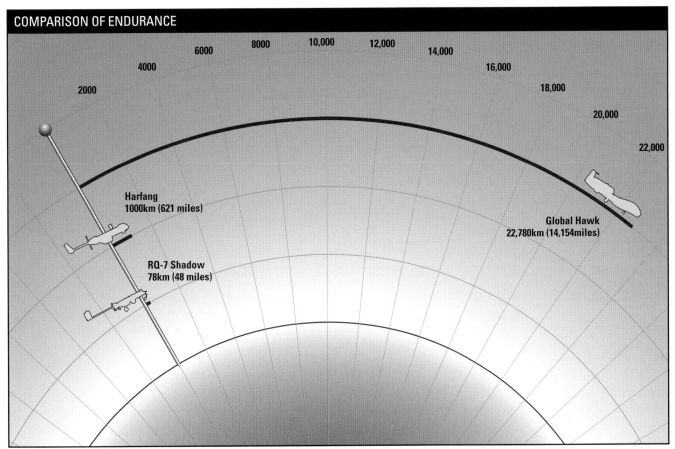

COMPARISON OF ENDURANCE

Harfang
1000km (621 miles)

RQ-7 Shadow
78km (48 miles)

Global Hawk
22,780km (14,154miles)

Above: The RQ-4A Global Hawk has come a long way since this maiden flight. The initial block of UAVs carried only IMINT (Image Intelligence) equipment, but some later models were upgraded with SIGINT (Signals Intelligence) and communications-relay equipment, significantly widening the capabilities of the aircraft.

Nevertheless, work was undertaken to create an extremely long-endurance UAV capable of carrying out persistent monitoring of an area and high-altitude surveillance of a combat zone. The approaches taken to these drones were very different in terms of capability and performance, therefore demonstrating

that there is often more than one way to solve a problem.

The problem, in this case, was the ability to put a camera or other instruments at very high altitude and keep it there long enough to reach the target area, and monitor that area over an extended period. Gravity is an unforgiving

opponent in this struggle – every kilogram has to be lifted and kept aloft, and that, in turn, means more engine power, more fuel for the engine and a heavier airframe to bear the added weight … which may, in turn, require yet more engine power.

The answer, for both high-performance UAVs and lighter craft, is to pare down weight, maximize lift for the power available and improve efficiency. The end results are quite different from one another, and have served to prove capabilities that may feed into the next generation of high-altitude UAVs.

RQ-4A Global Hawk

Global Hawk was designed from the outset to be an unarmed high-altitude reconnaissance platform, but that does not mean that the requirements it was created to meet remained consistent. Changing budgetary concerns, a strategic environment that continued

to evolve and the cost-cutting politics of defence procurement ensured that the specifications changed even as the project evolved to meet them.

The Global Hawk UAV is the first to venture into this environment, and, as such, it has had to cope with changing requirements and the need to prove technologies that may or may not have turned out to be blind alleys. Although a first-generation craft, it has matured as the project developed and did not have an entirely 'cold start'.

Conceived during the 1990s, Global Hawk was intended to be able to carry out the sort of missions a U-2 would be required for, and to do so more cheaply. It was also required to be capable of simultaneously carrying out electronic reconnaissance, i.e. detecting radar and signals emissions, as well as collecting radar, thermal and visual imagery.

As with many such projects, delivery

Above: The Global Hawk UAV has matured into a very capable system, by way of a complex and often confusing process. An initially fairly simple set of requirements was re-evaluated along the way, whilst the changing strategic and budgetary environments imposed new challenges on the designers.

of Global Hawk UAVs took place in several blocks, with each successive block incorporating new features or modifications. Block 0 consisted of pre-production aircraft, with the UAV going into service as the Block 10 model in 2003. Block 20 began production in 2006 and entered service in 2009.

Upgraded and modified versions continue to be developed. Enhanced signals intelligence capability was implemented for Block 30, with the Block 40 UAV selected for the multi-platform radar technology insertion programme

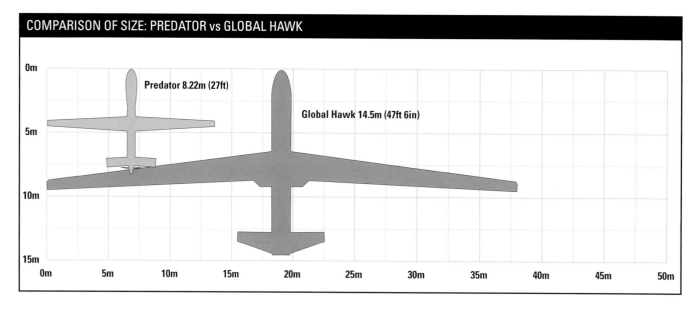

COMPARISON OF SIZE: PREDATOR vs GLOBAL HAWK

Predator 8.22m (27ft)

Global Hawk 14.5m (47ft 6in)

(MP-RTIP). This was originally to have both manned and unmanned platform involvement across NATO, improving ground surveillance capabilities, but the US has decided to go with a purely UAV programme. Future developments of the Global Hawk may incorporate longer range and additional sensor equipment, with a possibility that the UAV will eventually acquire a ballistic-missile warning capability.

Global Hawk has some visual similarities to a Predator or Avenger UAV, such as the angled tail section and the bulging frontal fuselage. It is, however, a much larger aircraft. Size and weight increased throughout the development process, from first flight in 1998 to current versions. Development will likely continue, perhaps with further increases.

In order to lift this fairly large load, Global Hawk uses a powerful turbofan engine. Unlike many UAVs, it can manoeuvre quite hard, even just after takeoff, and can climb steeply. It can reach altitudes of 18,288–19,812m (60–65,000ft) and remain aloft for 42 hours.

This long endurance and ability to cruise at fairly high speeds gives Global Hawk a very considerable range. In 2001, a Global Hawk made the first non-stop

crossing of the Pacific Ocean by an unmanned air vehicle. However, it is vulnerable to bad weather conditions and lacks de-icing equipment. Significantly, Global Hawk lacks the ability to see storms ahead of it and take avoiding action. Given its long flight duration, encountering bad weather is fairly likely, and, while aircraft such as the U-2 can fly 'above the weather', Global Hawk cannot always do this.

Despite these limitations, Global Hawk is a very potent sensor platform. Its optical and infrared sensors use a reflecting telescope, with thermal sensors and electro-optics capable of operating on a range of wavelengths. The synthetic aperture radar can also function as a ground target indicator. These systems are tied into the Global Hawk Integrated Sensor Suite (ISS), which can be augmented by additional sensors, such as MP-RTIP radar. This is an electronically scanned radar system, whose beam is steered electromagnetically rather than being fixed and aimed by the alignment of the UAV.

Survivability is enhanced by a stealthy design that minimizes radar return and thermal signature, and by the ability to fly above the engagement envelope of many

Right: Like many other large UAV projects, EuroHawk was intended to replace a manned reconnaissance platform – in this case, the Atlantique maritime surveillance aircraft. EuroHawk was based on Global Hawk, but was to carry additional SIGINT (Signals Intelligence) equipment. The programme encountered legislative troubles over UAV operations in civilian airspace.

ground-to-air weapons. Global Hawk also carries a radar warning receiver and an onboard jammer, and can deploy a towed decoy system.

Variants and overseas versions of the Global Hawk UAV have been created, among them EuroHawk. Developed for the German Air Force, EuroHawk was intended to meet the need for a long-range maritime surveillance aircraft, in which role its advanced Signals

Right: NASA uses the Global Hawk UAV for long-duration earth science missions. Missions include calibration of satellites and the development of new instruments, as well as measurement of conditions on the ground, at sea and in the atmosphere. Global Hawk UAVs are involved in research into how 'greenhouse gases' affect the earth's atmosphere.

Above: The MQ-4C Triton UAV was developed from Global Hawk for the Broad Area Maritime Surveillance role. Triton can monitor large areas of ocean with cameras, thermal imagers and instruments designed to detect radar and communications emissions. It is intended to replace the P-8 Poseidon maritime patrol aircraft.

SPECIFICATIONS: RQ-4A GLOBAL HAWK

Length: 14.5m (47ft 5in)
Wingspan: 39.8m (130ft 9in)
Height: 4.7m (15ft 4in)
Powerplant: Rolls-Royce North American F137-RR-100 turbofan engine
Maximum takeoff weight: 14,628kg (32,250lb)

Maximum speed: 570km/h (357mph)
Range: 22,632km (14,145 miles)
Ceiling: 18,288m (60,000ft)
Endurance: More than 34 hours

intelligence package and radar would have been very useful. EuroHawk ran into trouble over a need to ensure it was safe to operate in European airspace – Global Hawk has no detect-and-avoid system to protect against collisions with commercial and private aircraft. Plans have been mooted to modify the UAV to meet legislative standards, but the project remains in grave doubt.

Other potential users include Australia, Canada and Japan, all of whom are interested in the maritime surveillance role and possibly monitoring of Arctic regions. The US Navy has also ordered a variant, designated MQ-4C Triton. These UAVs have successfully competed exercises with the US and Australian navies, proving highly useful as wide-area surveillance assets. This is particularly valuable in open-ocean operations, such

DETECT-AND-AVOID SYSTEMS

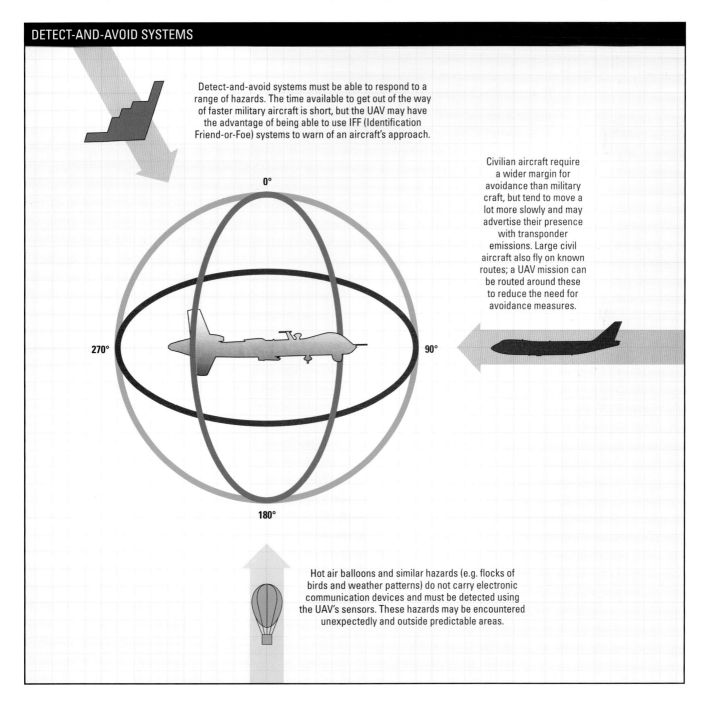

Detect-and-avoid systems must be able to respond to a range of hazards. The time available to get out of the way of faster military aircraft is short, but the UAV may have the advantage of being able to use IFF (Identification Friend-or-Foe) systems to warn of an aircraft's approach.

0°

Civilian aircraft require a wider margin for avoidance than military craft, but tend to move a lot more slowly and may advertise their presence with transponder emissions. Large civil aircraft also fly on known routes; a UAV mission can be routed around these to reduce the need for avoidance measures.

270°   90°

180°

Hot air balloons and similar hazards (e.g. flocks of birds and weather patterns) do not carry electronic communication devices and must be detected using the UAV's sensors. These hazards may be encountered unexpectedly and outside predictable areas.

as in the Pacific, where the sheer size of the region to be monitored requires a high-altitude, long-duration platform.

NASA has also taken a number of Global Hawks for experimental purposes. High-altitude aircraft and UAVs are extremely useful in researching upper-atmosphere and edge-of-space phenomena, and have applications in monitoring conditions on the ground or at sea as well. NASA's Global Hawks have been used for monitoring hurricanes and other weather phenomena, as well as the effects of pollutants on the atmosphere.

Zephyr

While Global Hawk took a fairly high-performance approach to creating a HALE (High-Altitude, Long-Endurance) UAV, Zephyr personifies an almost entirely opposite concept. Driven by two small propellers on the wings, Zephyr is a fragile-looking creation, composed mainly of a long wing with solar panels atop it. These are used to charge the UAV's batteries during daylight, enabling it to run its motors and systems at night.

Zephyr cruises at an extremely modest 55km/h (34mph), and climbs slowly to its operating altitude that is in excess of 21,000m (70,000ft). It

reaches about 12,192m (40,000ft) on the first day ,then completes its climb to altitude on the second. Zephyr is not by any means a fast-response platform, but it is not intended to be. Rather, the Zephyr drone is designed to remain on station for periods of up to three months, creating a persistent monitoring

Above: Zephyr is a flimsy-looking craft, but because it flies above the weather it does not need to contend with high winds other than when it is climbing to altitude or returning to earth. Reaching operating altitude requires several hours of slow ascent, so a calm launch window is essential to success.

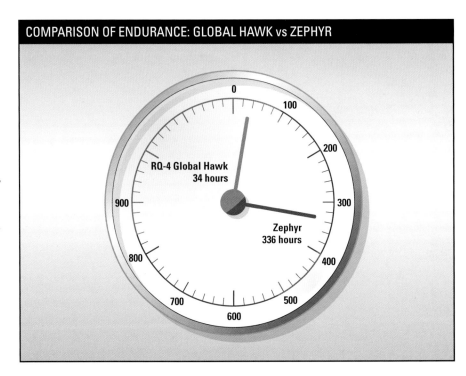

COMPARISON OF ENDURANCE: GLOBAL HAWK vs ZEPHYR

RQ-4 Global Hawk
34 hours

Zephyr
336 hours

capability that can be retasked to a new area, but not immediately.

Until recently, the technology to create such a craft did not exist. The primary challenges in creating a UAV of this sort were keeping weight as low as possible and providing (and storing) sufficient power for its motors and systems. The small electric motors that drive Zephyr's propellors do not require much power, but motors and batteries are relatively heavy. Conversely, Zephyr requires no fuel, which not only reduces the weight that must be carried, but also removes any need to return to base for refuelling. In theory, so long as it gets enough solar energy to stay above 15,240m (50,000ft), Zephyr can stay aloft more or less indefinitely.

The frame uses carbon fibre and is shaped to create a low-drag, high-lift configuration that further minimizes the amount of power required to keep Zephyr in the air. Such a lightweight frame cannot resist particularly bad weather, but Zephyr's operating altitude is well above weather effects, so this is not a problem once it reaches its target altitude. Ascent will require a fairly long 'window' of

clear air, so launch times are somewhat constrained by ground conditions, but this is offset by the ability to stay up for months once launched.

Launching requires five people and a certain amount of teamwork. The launch team holds the craft aloft and runs into the wind with it, giving Zephyr enough initial airspeed to provide lift. Its engines then accelerate gently and it begins its long climb to altitude. Whlle somewhat unglamorous, this mode of launch places little strain on the fairly fragile airframe and does away with any need for undercarriage. On return, the drone lands on its belly and is light enough to come down at very low speed without stalling, thus reducing the probability of damage.

The primary advantage of an extremely long-endurance UAV of this sort is the ability to monitor an area for long periods for very little cost. It is not possible to carry extremely high-capability instrument

packages – and weapons are, of course, out of the question – but a steady trickle of information from a persistent source can be extremely useful.

Zephyr's instrumentation or other payload package is carried in a removable pod. This usually contains a lightweight electro-optical system, but can also carry a communications relay unit. Other payloads are under development for a variety of purposes.

Zephyr has already broken several world records, including endurance and height records for an unmanned vehicle. It has been incorporated into the High-Altitude Pseudo-Satellite (HAPS) programme, which investigates the use of drones as low-cost alternatives to orbital satellites. A high-altitude drone can act as a communications relay device to allow signals to be 'bounced' over the horizon in a similar manner as a satellite, but it is vastly cheaper and simpler to build.

Below: Where Global Hawk hurtles into the air like a fighter, Zephyr is launched by a team of people who run with it until it gains enough speed to start ever-so-gently climbing. Whilst this may seem a little amateurish, it makes possible the incredible endurance and height performance of this advanced UAV.

Long-endurance reconnaissance drones

As with any military technology, UAV designers have to balance capability against cost and the risks inherent in implementing a new technology or device. A UAV built around a potentially excellent system that fails to deliver the expected performance could lose its developers a fortune if orders fail to materialize. Similarly, extremely high-performance UAVs may not be affordable to many users, either in terms of obtaining enough to provide suitable coverage or, in some cases, a single example might be too expensive to justify purchase, whatever capabilities it might have.

Above: Information handling and distribution is an essential part of military operations. Commanders at different levels have different needs, which can often all be served by data filtered out of the same raw feeds. UAVs have created new information-gathering opportunities, but the technology for handling that information must keep pace.

Many militaries take a high-end/low-end approach to procurement, obtaining a few expensive high-capability systems and a larger number with more basic capabilities to ensure that there is adequate coverage. A modest amount of drone-gathered information available to many units on the ground may often be worth more than truly excellent capability in just one place.

The various long-endurance reconnaissance drones that are available can provide reconnaissance of value at both the strategic and tactical levels. A company commander on the ground needs to know what is happening imminently in his immediate vicinity, whereas the planning staff of his parent division may require a wider picture that will give an indication of enemy capabilities and intentions over a longer period. Both can benefit from information gathered by the same UAVs, provided it is well handled and distributed timely.

Thus personnel involved in intelligence operations have had to evolve procedures to keep pace with new technological developments. Too much information can be as much of a problem as too little

– it does not matter whether an enemy force is missed in the clutter of data overload or simply not seen at all. For this reason, a reconnaissance drone cannot be considered in isolation. Information-handling technology at the UAV's control station is every bit as important as the sensors it carries. A drone is part of a package that offers great benefits, but only if all components are well used and personnel understand how to get the most out of the data gathered.

Firebee

The Ryan Firebee started out as a target drone in the 1950s. It was developed in response to a need for a jet-powered target vehicle that could simulate fast aircraft and missiles for training purposes. The Firebee was intended from the outset to be reusable and to carry electronics. It had scoring systems aboard and flares at the wingtips to attract infrared-homing missiles away from the fuselage.

Firebee drones were designed to be reconfigured to mimic various possible opponents, including the ability to deploy countermeasures and to fly extremely low and fast in the manner of a high-speed bomber making a 'dash' penetration of hostile airspace. After being successfully attacked, the drone would deploy a parachute, and either be caught in flight by a helicopter or float in water until retrieved. Many Firebees had a long career with many successful recoveries and were decorated with symbols denoting an air or sea retrieval.

In the 1960s, it became increasingly obvious that the Firebee UAV could be used for high-altitude strategic reconnaissance purposes. It was modified to reduce radar signature, using knowledge gained in part from its own many interceptions and shoot-downs. These modifications were rather basic by modern standards, consisting mainly of a remodelled air intake screen to reduce its radar reflections, and a combination of paint and radar-absorbent materials

Above: The Firebee UAV originally entered service in 1951 as a target drone. It was designed to be dropped from an underwing launcher or to take off from the ground using a RATO (Rocket Assisted Takeoff) booster. Service as an information-gathering platform began in the 1960s.

coating the external surfaces. The reconnaissance version of the Firebee was capable of loitering at high altitude – up to 22,860m (75,000ft) for several hours after launch from a C-130 control aircraft. It could be directly commanded, or set to fly a predetermined course before leaving the target area for retrieval. Firebees were used to conduct signals intelligence and communications intercept work over North Korea, among other deployments, during a long career.

The Firebee UAV was upgraded with better engines and new capabilities throughout its career, and many continued to be used as target drones, while some examples were serving as reconnaissance platforms. Other uses were eventually found, such as deploying chaff to confuse Iraqi air defence radars ahead of air strikes in 2003. This concept was nothing new – it was first done during World War II – but previously it required aircraft to enter defended airspace. Sending drones to drop chaff greatly reduced the risk to manned platforms.

The Firebee UAV is normally operated from a C-130 aircraft, which carries the drones on underwing pylons. It can also be launched from the ground using an auxiliary booster. Modern Firebees use GPS guidance and a variety of countermeasures to simulate various targets, and have also been put to use for tasks such as damage assessment during exercises.

RQ-170 Sentinel

Since first flying in 2007, the RQ-170 Sentinel has been built in relatively small

Left: The Firebee drone was initially designed to be air-launched from under the wings of an A-26 Invader. Second-generation Firebees, developed in the 1960s and designated Q-2C or BQM-34A, were instead launched from underwing pylons on a C-130 Hercules. The parent aircraft can launch and control up to four Firebees.

numbers – perhaps 20 or so examples – and has not been widely deployed. It is a dedicated ISTAR (Intelligence, Surveillance, Target Acquisition, Reconnaissance) platform that has also been used for damage assessment following airstrikes.

The RQ-170 Sentinel bears a distinct resemblance to manned stealth aircraft, notably the B2 Spirit bomber and the now-retired F 117 Nighthawk 'stealth fighter'. Its 'flying wing' shape is designed for low-observability, enabling it to undertake reconnaissance missions on heavily defended airspace, or in a political environment where detection might result in an unwanted incident.

Much of the aircraft's structure is formed from lightweight composite materials, which also reduce radar return compared to metals of equivalent strength. It is powered by a turbofan engine that probably gives a high thrust-to-weight ratio, although performance details are not widely available. Likewise, service ceiling is not known, but is probably somewhere in the medium-altitude region with a maximum of about 15,240m (50,000ft). Endurance is undoubtedly a significant factor, but it also remains unknown.

Sentinel is not a weapons platform; it was designed from the outset for covert information gathering. It carries an electro-optical camera in the underside of the nose, and thermal as well as electro-optical sensors in the upper surfaces of its wings. Synthetic aperture radar and an electronically scanned array radar system are carried in the belly fairing. The craft may also carry electronic warfare and signals intelligence payloads and/or radioactive-particle detectors for use in locating nuclear weapons facilities.

The stealthy nature of this craft, and its possible ability to detect a nuclear weapons programme, led to speculation that the RQ-170 Sentinel deployed to Afghanistan was, in fact, being used to monitor Iran instead.

Above: An RQ-170 crashed in Iran during late 2011, probably due to systems failure. The vehicle was paraded by the Iranian Government, which claimed credit for downing it, either by gunfire or some kind of control interference. This, like the claim that the UAV has been reverse-engineered, is very probably spurious.

There was no real need for a stealthy drone over Afghanistan due to the lack of sophisticated radar systems in the hands of the Taliban, although that does not rule out the possibility of using the drone to gain operational experience, or simply because it was available.

Sentinel UAVs began operations in Afghanistan in 2007, which were openly acknowledged in 2009. It has also served in South Korea, replacing U-2 reconnaissance aircraft for missions monitoring North Korean activity. The loss of an RQ-170 over Iran proved that reconnaissance overflights of that country were being undertaken while the Sentinel UAV was based in Afghanistan, but this does not mean that was the

Left: The RQ-170 Sentinel UAV remains shrouded in secrecy, but appears to be a 'stealth' aircraft equipped for signals and image intelligence operations. An RQ-170 reportedly provided surveillance information on the compound used by Osama bin Laden, and supported the operation to eliminate him by monitoring local radio transmissions.

exclusive mission of that drone. Indeed, data gathered by Sentinel was used in planning the attack that killed al-Qaeda leader Osama bin Laden.

The UAV is controlled from a ground station, but can operate autonomously for most of a mission. While operational losses may be inevitable with any aircraft, the prospect is minimized by redundant and robust electronics and a return-to-base system. The Sentinel UAV can take off and land autonomously, and will do so when contact is lost, finding its own way home if at all possible. Only a severe malfunction or enemy action would cause it to crash.

Iranian officials have claimed that they have reverse-engineered the RQ-170 that crashed in Iran and have created their own version, which may carry weapons. This seems highly unlikely, as this is a very complex craft, and anyone capable of learning its secrets and implementing an improved version in such a short period of time would probably not need to – they would already be flying something better.

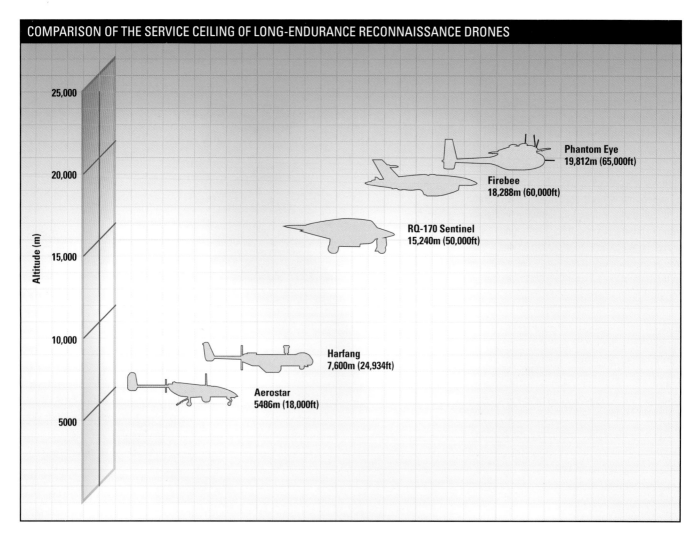

COMPARISON OF THE SERVICE CEILING OF LONG-ENDURANCE RECONNAISSANCE DRONES

Altitude (m)

25,000

20,000

15,000

10,000

5000

Phantom Eye
19,812m (65,000ft)

Firebee
18,288m (60,000ft)

RQ-170 Sentinel
15,240m (50,000ft)

Harfang
7,600m (24,934ft)

Aerostar
5486m (18,000ft)

Aerostar

Aerostar uses a twin-tail, pusher propeller configuration favoured by several UAV manufacturers. This allows cameras to be mounted in the nose of the UAV and a compartmentalized configuration to be used, with the payload bay ahead of the motive section. It was introduced in 2000, and was quickly adopted by the Israeli Defence Force (IDF) for tactical reconnaissance, border patrol and security operations. Other users followed suit, and the Aerostar UAV has also undertaken trials with the Israeli police.

The use of helicopters for traffic surveillance or tracking of suspect vehicles is long established; a UAV allows the same capabilities at a lower cost, and will be able to stay on station far longer than a manned helicopter. Since a small drone undertaking routine traffic surveillance is unlikely to be spotted, drivers cannot be sure they are not under observation at any given time. It has been suggested that this will deter many drivers from speeding and committing other traffic offences.

There are those who think that using drones in this manner is somehow cheating, as the driver cannot know when he has a good chance of getting away with breaking the law, but Israeli law enforcement officials would seem to disagree. During a trial operation in which police cars equipped with an Aerostar control unit were deployed, numerous motorists were stopped and shown video of themselves committing various offences. The project was sufficiently promising from the point of view both of driver education and of prosecution of offences, that widespread implementation of UAV traffic surveillance is now being contemplated.

Aerostar was developed with interoperability and flexibility in mind. The latter is important when seeking export customers, as the end user may have a particular sensor system or equipment package in mind when they are selecting a drone to carry it. A successful export

Right: Aerostar uses an integrated flight control system, which handles navigation, engine output, heading and payload management during all operations, including takeoff and landing. The operator gives general instructions, telling the UAV where to go and what to look at, but need not be concerned with details.

Below: UAVs, such as Aerostar, which use a conventional aircraft-style layout, are easier to develop than advanced, stealthy, flying-wing versions. A small-aircraft design with a pusher propeller offers a good payload-to-weight ratio, and creates maximum space in the payload section whilst not impeding forward-looking instruments.

design must be able to accommodate a range of packages.

Not all users are so concerned with interoperability; some will prefer to use their drones in isolation. For example, Aerostar has been used to test detect-and-avoid systems that are planned for implementation aboard other UAVs, such as Predator or Reaper. This application

does not need integration with other systems. However, for those that favour a network-centric approach, a UAV that can be tied in with helicopters, warships and other platforms offers some significant advantages.

Like many reconnaissance UAVs, Aerostar is designed to feed its data back to the command station in real time,

from where it can be disseminated as necessary. The command station can take the form of a multi-drone operations centre or portable payload control station, which allows personnel close to a target area to direct the information-gathering process of a UAV close overhead to obtain precisely the information they require at the time. This enhances

AEROSTAR EXTENDED OPERATING RADIUS

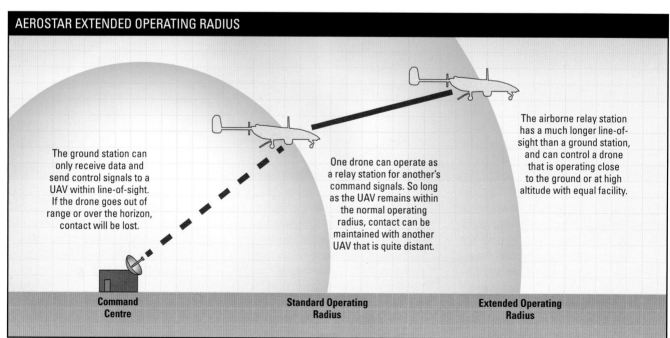

The ground station can only receive data and send control signals to a UAV within line-of-sight. If the drone goes out of range or over the horizon, contact will be lost.

One drone can operate as a relay station for another's command signals. So long as the UAV remains within the normal operating radius, contact can be maintained with another UAV that is quite distant.

The airborne relay station has a much longer line-of-sight than a ground station, and can control a drone that is operating close to the ground or at high altitude with equal facility.

Command Centre

Standard Operating Radius

Extended Operating Radius

Left: The Aerolight UAV is designed to be quickly assembled and launched from either a catapult or a conventional runway. Takeoff and landing are automated, whilst once in flight the operator can choose whether to guide the UAV manually or to let it operate autonomously.

mission-stage planning and allows reconnaissance of an area just before a ground force moves into it.

The UAV itself is designed to be as simple to operate and as mistake-proof as possible. Critical commands are queried by the control station to give

the operator a chance to confirm or abort a potentially risky manoeuvre, but most navigation is autonomous, using GPS and inertial navigation. In addition to thermal and electro-optical sensors, payloads include synthetic aperture radar and electronic warfare or intelligence packages. Multiple packages can be carried when required, giving a wide range of capabilities during a single mission.

It is also possible to extend the Aerostar's operating radius by using one Aerostar drone as a relay for command signals sent to another. Since control range is limited by line-of-sight to the ground station, this gives operators the ability to control a drone 'over the horizon' without using a satellite link. The directional antennae aboard the UAV can 'see' another drone operating at high altitude at far greater distances than contact can be maintained with a ground antenna. This relay system does away with the need to operate at higher altitudes as distance from base increases; relayed signals can be received by a

drone that has come down low to bring its cameras to bear on a target that cannot otherwise be investigated.

Heron/Harfang

First flying in 1994, the Heron UAV uses a twin-tail configuration and pusher propeller. It has seen significant export success, having been taken into service with numerous users worldwide. Designed as a Medium Altitude Long-Endurance (MALE) UAV, Heron carries a range of mission packages, including signals intelligence and communications intercept equipment, thermal and electro-optical cameras and radar systems. It has been used for missions ranging

Below: During 2008 and 2009, the Israeli military successfully experimented with the integration of Heron UAVs at the brigade level for situational awareness and battlefield surveillance. By placing the UAV squadrons relatively low in the chain of command, the Israelis achieved more timely dissemination of information to those needing it most.

SPECIFICATIONS: AEROSTAR

Length: 4.5m (15ft)
Wingspan: 8.5m (28ft)
Height: 1.3m (4ft 3in)
Powerplant: 1 x Zanzottrra 498i 2-stroke boxer engine
Maximum takeoff weight: 220kg (485lb)
Maximum speed: 200km/h (126mph)
Ceiling: 5486m (18,000ft)
Range: 250km (155 miles)
Endurance: 12 hours

Above: The Harfang UAV started out as an attempt by the French military to plug a procurement gap by buying an off-the-shelf UAV (Heron) from Israel. Modifications to the Heron drone created Harfang, which has achieved success in roles as diverse as providing security for a papal visit and military reconnaissance over Libya.

SPECIFICATIONS: HARFANG

Length: 9.3m (30ft 5in)
Width: 16.6m (54ft 5in)
Powerplant: 1 x Rotax 914F turbocharged engine
Maximum takeoff weight: 1250kg (2756lb)
Maximum speed: 207km/h (129mph)
Ceiling: 7600m (24,934ft)
Range: 1000km (621 miles)

from battlefield surveillance and missile or rocket attack warning to artillery fire adjustment.

Heron can be operated directly from the ground station using either line-of-sight radio commands or a satellite interface, or can operate autonomously. Its mission package, likewise, can be controlled directly or can operate according to preset parameters. Navigation is by GPS with an automated takeoff and landing system, and, in the event of control signal loss, the UAV is programmed to return to base and to land itself.

Heron drones have served in Afghanistan and other conflict zones, beginning with Gaza. Israeli Defence Force operations in Gaza in 2008 and 2009 made extensive use of Heron and other UAVs for tactical reconnaissance and battlefield surveillance. The operation was characterized by close cooperation between forces on the ground and their supports, including artillery, air and naval units. This included the rapid passing of information and reconnaissance data between formations from different arms of service, enabling rapid response and effective close support in a very complex and cluttered battlespace.

While numerous nations have chosen to buy Heron drones as offered, France instead elected to develop a derivative named Harfang. This UAV was developed to replace the force of RQ-5 Hunters, which were, at that time serving with the French military. It first flew in 2006 and entered service in 2008. Since then, Harfang has supported French operations in Afghanistan, Mali and Libya, and was deployed as a security measure during the Pope's visit to France in 2007.

Meanwhile, a new version of Heron, named Heron TP or Eitan, has been developed. Eitan adds new capabilities to Heron's repertoire, including air-to-air refuelling. It can operate at higher altitudes and has a de-icing system to facilitate operations in cold air and at altitude. Eitan has several bays and attachment points for payload items, enabling it to be optimized for various missions – some systems require different placement aboard a UAV than others.

The Eitan UAV also has a more powerful engine and improved avionics over its predecessor. It can fly above the altitudes used by commercial traffic

– up to 12,192m (40,000ft) – and is capable of mission durations of up to 36 hours. This may be extended by buddy-refuelling from another Eitan drone, creating the possibility of extremely long-duration flights.

Below: The Heron TP UAV is sometimes designated Heron 2 or Eitan by its Israeli operators. In addition to the usual reconnaissance and surveillance tasks, Heron TP was developed to provide strategic missile defence and airborne refuelling capabilities. Its ability to operate at high altitude places it beyond the interception capabilities of many opponents.

Above: Phantom Eye was the first UAV to be powered by liquid hydrogen, benefiting from excellent fuel economy. This permits Phantom Eye to maintain a high altitude for four days at a time. Experience with this craft is being put to use developing a next-generation hydrogen-fuelled drone with an endurance of over a week.

Phantom Eye

First flying in 2012, Phantom Eye was the first UAV to make use of liquid hydrogen fuel. Its twin engines are derived from more conventional ground vehicle engines, with turbochargers to enable them to function in an oxygen-poor environment, such as that encountered at high altitude. This system is notable for its low carbon footprint, which may seem to be a strange consideration for a military system. However, in recent years environmental concerns have begun to influence which military projects get funding and which do not; this may become an increasingly important factor in defence procurement in future.

More immediately, the hydrogen-powered engines are fuel efficient, which is of critical importance when building a drone for persistent intelligence gathering and similar long-duration operations. Phantom Eye is capable of carrying out a range of missions, not all of which are military. It could be used for environmental monitoring or other scientific applications by swapping its payload. It has been put forward as a communications relay platform, using relays of drones to provide constant coverage.

The challenges inherent in getting piston engines to work at extreme altitudes were not small. Before installation in an operational UAV, the engines were extensively tested and

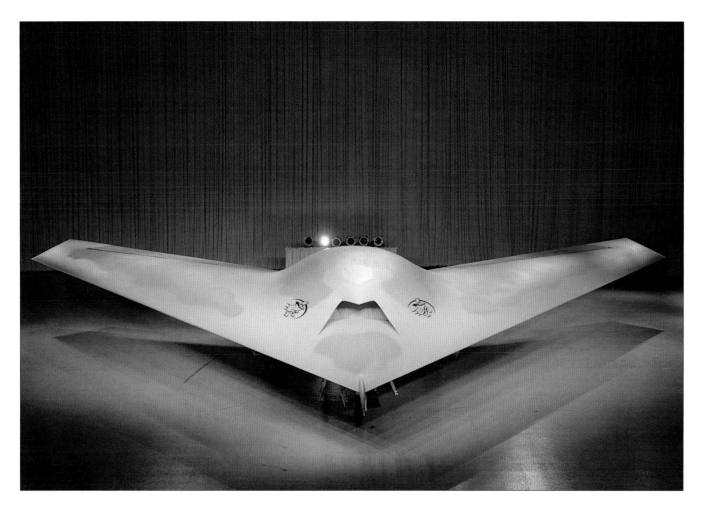

Above: Phantom Ray is a technology demonstrator based on the earlier X-45 project (a candidate for the US Navy's Unmanned Carrier-Launched Surveillance and Strike programme) and concepts proven with Phantom Eye. It offers greater lifting capacity and performance over Phantom Eye, extending endurance to 10 days or more.

refined in ground-based facilities that simulated conditions at high altitude. The engines did not merely have to demonstrate their ability to tolerate thin air and cold conditions, they had to be able to do so for an extended period.

Phantom Eye's overall design is intended to be efficient rather than stealthy. It can operate at very high altitudes – around 19,812m (65,000ft) – where interception is very unlikely. With this in mind, the designers concentrated mainly on payload capacity, altitude and loiter time. They were able to draw on experience with Condor, an earlier UAV project that set a new height record for piston-engined aircraft. Condor was also the first UAV to make a completely autonomous flight.

Designed to be operated manually if necessary, Phantom Eye can take off, land and navigate autonomously, and it will make an automatic safe landing in the event that control with it is lost. Contact with the ground station is made by way of a satellite link, allowing real-time data from the UAV to be collected by the ground station and disseminated as necessary.

Phantom Eye's sensor package includes electro-optical and infrared sensors for the reconnaissance role, but it can carry a range of equipment

packages. Its ability to remain on station can be extended further by the addition of large-capacity fuel tanks. The UAV's standard capability is reported at about four days, but the designers hope to push this to seven to 10 days or more as the design matures.

An enlarged version, named Phantom Ray, first flew in 2011. Phantom Ray was designed to increase endurance and also the lifting capacity. It has been put forward as a testbed for high-altitude technologies and other UAV-borne systems.

Medium-range reconnaissance drones

Medium-range drones trade absolute capability for relatively small size. This makes them more portable and harder for hostiles to detect, but does not always mean they are of low utility. It is true that a smaller UAV has less lifting capability and less space for payload, so it cannot carry such a wide range of equipment as a larger drone, but, as technology advances, more can be done in a small space.

Above: The Fury 1500 UAV can be launched in a small space using its pneumatic catapult, enabling it to operate in cluttered terrain, close to water, or from a ship. Its recovery system requires little space and makes operations possible from a small vessel.

One solution to the limited space available aboard a medium-sized drone is to use a modular payload and swap sensor and equipment packages as the mission requires. Alternatively, a single sensor package can be integrated into the drone's structure. This is not as flexible, but, with no need to connect other equipment or access the sensor package for removal and replacement, it can be fitted into the space available in the most efficient manner possible.

Another problem for designers of smaller drones is the ability to provide enough power for some onboard systems. Cameras do not require much energy, but radar and communications equipment tends to be significantly more power-hungry. A drone that runs entirely on batteries cannot power such systems along with its motors for very long, whereas one that uses a combustion engine can generate power in flight and can operate all its systems until it runs out of fuel.

Thus there is an optimum size for a medium-range UAV: large enough to carry a useable payload, but not so big as to become difficult to transport and launch. Designers may be tempted to increase the size of their drone just a little more to fit some desirable component or add extra capability, but there is always the danger that by doing so they may make the drone too large or too costly for its intended role.

Fury 1500

Fury 1500 is a 'runway independent air vehicle'. It is launched from a pneumatic catapult and is recovered using a net. While this is not as glamorous as some aircraft-type drones, it permits use aboard small ships and in other confined areas that would otherwise make UAV operations impractical.

Fury 1500 uses an advanced delta-wing configuration with a three-bladed propeller at the rear that is driven by a heavy-fuel engine. This generates very considerable amounts of electrical power for onboard systems – Fury 1500 is

marketed as 'best in class' for onboard power generation.

The Fury 1500 UAV is equipped with thermal and electro-optical cameras, synthetic aperture radar, plus electronic, signals and communications intelligence equipment, enabling it to carry out a range of intelligence-gathering missions. Its electronics are shielded to protect them from electromagnetic effects such as jamming or other strong radio-frequency interference. Payload systems are designed on a 'plug and play' basis, allowing other packages to be swapped in as needed.

Control is via a line-of-sight data link or satellite communications to a ground station that handles all processing and dissemination of data from the UAV. It can reach altitudes of 4570m (15,000ft) and has a ferry range of over 2700km (1680 miles). This is not an operating radius – it is the maximum distance the

drone can fly in a straight line. Operating radius would be less than half this figure, assuming some loiter time to carry out its surveillance or reconnaissance mission. Maximum mission duration is around 15 hours of flight time.

Falco

The Falco UAV was developed in Italy with the needs of the Pakistan Government in mind. With huge areas of at times inhospitable and underdeveloped land to cover, Pakistan faces significant homeland security challenges. Given its location between Afghanistan and the Middle East, it is inevitable that this nation is affected by the current turbulent times. Lacking adequate resources to patrol vast territories on the ground, drones may be a cost-effective solution.

Falco is currently a reconnaissance UAV only, but it has a single hardpoint under each wing with the capability to

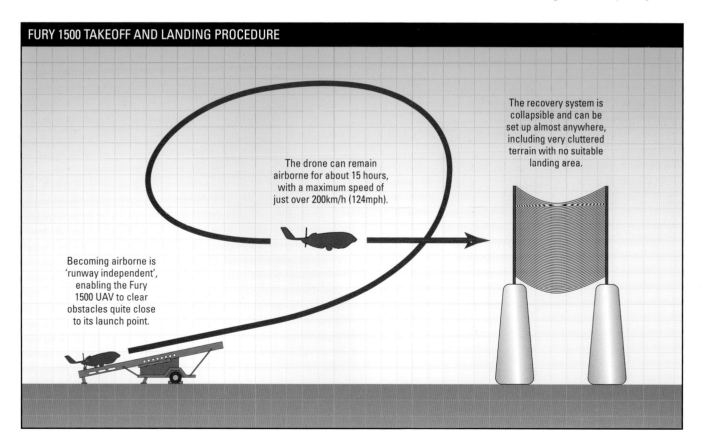

FURY 1500 TAKEOFF AND LANDING PROCEDURE

The recovery system is collapsible and can be set up almost anywhere, including very cluttered terrain with no suitable landing area.

The drone can remain airborne for about 15 hours, with a maximum speed of just over 200km/h (124mph).

Becoming airborne is 'runway independent', enabling the Fury 1500 UAV to clear obstacles quite close to its launch point.

Below: Falco's two underwing hardpoints may enable it to carry small weapons with a total weight of 25kg (55lb). This rather modest strike capability is better suited to small-scale operations against insurgents than large-scale warfighting, but would give Falco's operators a useful extra capability in counter-insurgency operations.

carry a small missile. These will probably be laser-guided precision weapons, giving Falco a limited strike capability. In the meantime this UAV is put forward as a military reconnaissance platform and security asset, with applications ranging from border patrol to fishery protection or suppression of smugglers.

Falco's sensor package includes thermal sensors and electro-optical cameras, a laser designator and an NBC sensor that can detect the use of weapons of mass destruction or harmful chemicals in a target area. It can also carry synthetic aperture radar or maritime surveillance radar, plus an array of

NBC SENSORS SYSTEM ONBOARD FALCO

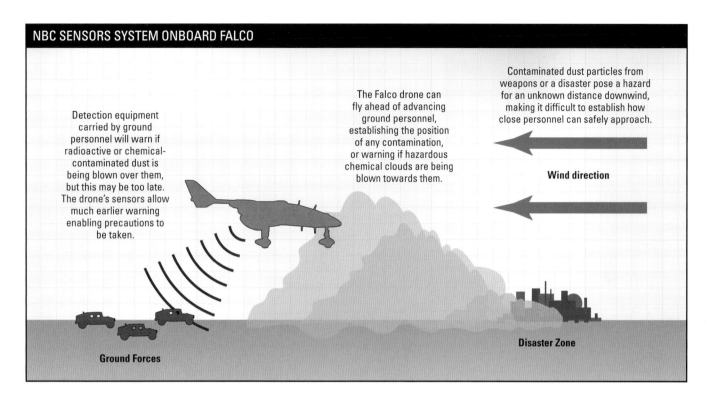

Detection equipment carried by ground personnel will warn if radioactive or chemical-contaminated dust is being blown over them, but this may be too late. The drone's sensors allow much earlier warning enabling precautions to be taken.

The Falco drone can fly ahead of advancing ground personnel, establishing the position of any contamination, or warning if hazardous chemical clouds are being blown towards them.

Contaminated dust particles from weapons or a disaster pose a hazard for an unknown distance downwind, making it difficult to establish how close personnel can safely approach.

Wind direction

Disaster Zone

Ground Forces

electronic warfare equipment.

Designed for operations in conditions over Pakistan, Falco was nevertheless tested in a range of environments, including very cold and damp northern European climates. It is capable of making a short-run takeoff from a primitive airstrip, or can be launched using a pneumatic catapult. This makes operations in fairly cluttered terrain less problematic. Recovery can be by way of a conventional wheeled landing controlled automatically by the UAV's internal systems. Alternatively a 'tactical short landing' mode can be implemented or – if there is insufficient space for even this – the UAV can deploy a parachute and

Left: Falco can take off autonomously and is equipped for short-field operations in primitive conditions. Landing is normally conventional, but, at need, the UAV can parachute itself back to the ground. The UAV's undercarriage and general design are robust enough to survive rough landings without damaging the sensitive payload.

Below: The Falco ground station receives near-real-time data from the UAV and can pass information to commanders in the form of video clips, still images and sensor readouts. The ground station can be used as a training simulator or for pre-mission planning.

SPECIFICATIONS: FALCO

Length: 5.25m (17ft 2in)
Width: 7.2m (23ft 6in)
Height: 1.8m (5ft 9in)
Powerplant: 1 x petrol
Maximum takeoff weight: 420kg (926lb)
Maximum speed: 216km/h (134mph)
Ceiling: 6500m (21,325ft)
Endurance: 14 hours

come back to earth vertically.

Deployment is normally in units of four aircraft plus a control station and a data-handling terminal. As with many drone control stations, Falco's control system has a simulator mode for training or mission planning. Control is line-of-sight, limiting the distance the UAV can operate from its control suite to about

200km (125 miles). With a mission duration of up to 14 hours, this enables persistent surveillance of an area using a relay of drones, and control range can be extended if necessary using a relay, or by handing off control of the UAV to another ground control station.

The Falco drone is the first UAV to be used by United Nations' peacekeeping

COMPARISON OF THE ENDURANCE OF MEDIUM RANGE RECONNAISSANCE DRONES

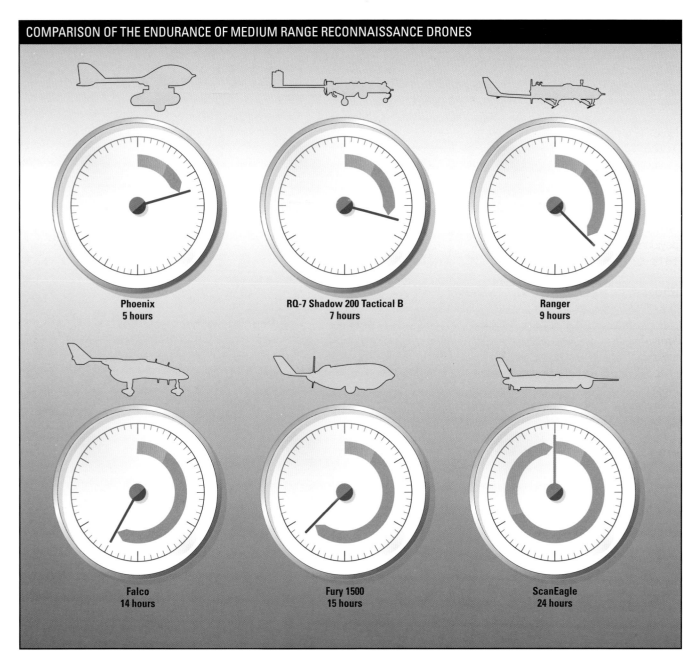

Phoenix
5 hours

RQ-7 Shadow 200 Tactical B
7 hours

Ranger
9 hours

Falco
14 hours

Fury 1500
15 hours

ScanEagle
24 hours

Left: The Phoenix UAV was not a huge success. Its convoluted design required landing upside-down to protect the underslung sensor equipment. Although the official verdict was that this UAV proved very useful, personnel who worked with Phoenix have a variety of unpleasant things to say about it.

Below: Resembling a World War I aircraft in many ways, Ranger uses a space-efficient design and incorporates a number of 'stealth' features. It is designed to land on grass, ice or a flat space such as a road, or if nowhere suitable is available it can make a parachute landing instead.

forces. It was deployed to assist monitoring of militias operating in the Democratic Republic of the Congo as part of a programme intended to reduce conflict in the region. An expanded version, likely to have strike capability, is named Falco EVO. It can carry a larger payload and extends mission times to 18 hours or more.

Phoenix

The Phoenix UAV first flew in 1986 and saw active service with the British Army. It used a fairly typical twin-boom tail configuration, but, unusually, was driven by a 'tractor' propeller. Placing the propeller at the front of the fuselage created a mounting problem for forward-facing cameras, which was solved by carrying them in a pod under the centre of the fuselage. This, however, made the cameras prone to damage during landing. It also ruled out a rolling takeoff from any sort of runway.

The takeoff problem was solved by using a catapult to launch the UAV along a rail mounted on a truck, but landing was more of an issue. The solution was to deploy a parachute that would cause the drone to land on its back, and to add a humped area to the dorsal surface that would absorb the impact of a landing. While undignified, this system allows

the Phoenix UAV to operate in a very cluttered environment. This was a design requirement, since, at the time, the British Army considered it probable that its next major conflict would be in the urbanized terrain of northern Europe.

In practice, the Phoenix UAV did not enter service until 1999, going on to serve in Kosovo and later Iraq. There, it suffered significant losses, although

many were due to operators deliberately keeping their drones over the target area until their fuel ran out. The decision to trade a potentially reusable drone for a few minutes' more data was a difficult one, but the British Army clearly thought it worth the sacrifice.

Equipped with an infrared camera and with an endurance of four or five hours and a ceiling of 2745m (9000ft), Phoenix

Above: Ranger was the first fixed-wing (i.e. aircraft-type) UAV to receive certification to operate in civilian airspace. In addition to its military role, the Ranger UAV is intended for use in security operations, and has a range of disaster-management applications, including the monitoring of volcanoes and earthquakes.

offered only a modest performance. Whille the British Army officially claims that this UAV was extremely valuable for artillery spotting and reconnaissance, many of those who worked with it are less complimentary. Phoenix was retired in 2006 and was replaced with the rather better Desert Hawk UAV.

It has been suggested that by the time Phoenix was retired, virtually all surviving examples were unserviceable. Apart from those deliberately sacrificed, a large proportion of Phoenix drones were shot down, crashed due to faults, or simply went missing due to unknown circumstances. This depressing tendency to not come back led to a poor reputation among Phoenix' users and its inclusion in some engineering courses as an example of how not to approach a project.

Ranger

A joint Swiss-Israeli project developed to meet the needs of the Swiss Air Force, Ranger entered service in 1999 and has

since been adopted by Finland as well. It drew on an earlier UAV named Scout created by Israeli designers. Although not shaped like a 'stealth' aircraft, Ranger is small and uses composite materials to reduce radar return.

The Ranger UAV consists of a twin-boom tail section attached to the straight wings, with a blocky fuselage resting atop the wings. A pusher propeller, driven by a two-stroke internal combustion engine, is located at the rear of the fuselage, with the forward area available for payload. As standard, this includes electro-optical and forward-looking infrared systems.

A retractable turret contains additional cameras and thermal sensors, while the fuselage can carry a synthetic aperture radar and/or electronic and signals intelligence equipment. An emergency parachute is also carried, enabling the drone to be brought down safely if there is no suitable landing area, or in the event of a malfunction. More commonly, Ranger lands on skids that enable it to

use grass or ice as a landing area, or any road or similar surface. The propeller is mounted high and kept clear of the ground during landing. A rolling takeoff is not possible, of course – Ranger uses a catapult for launch, creating a minimal takeoff distance.

Ranger is a multi-mission platform intended to provide reconnaissance, electronic warfare, electronic and signals intelligence and artillery spotting functions, and additionally has non-military applications. Its sensors can detect nuclear radiation and could be used to provide information on disasters, such as floods, fires and earthquakes.

The Ranger UAV is controlled from a mobile ground station mounted on a truck, using a line-of-sight radio link out to a range of about 180km (112 miles). Data is transmitted in real time to the remote communication terminal of the ground station.

RQ-7 Shadow

The RQ-7 Shadow UAV is used for tactical reconnaissance, target acquisition, damage assessment and battlefield awareness at the brigade level by the US Army. It was also adopted by

Above: An RQ-7 Shadow system consists of three UAVs plus a fourth disassembled as a spare. A unit consists of two light vehicles carrying the UAVs and launchers, and two more with trailers providing personnel mobility. This unit provides 72 hours of operations before needing resupply.

the US Marine Corps and the Australian and Swedish Armies. RQ-7 has served in Afghanistan and Iraq since its introduction in 2001, with all original RQ-7A models now retired and replaced by the more capable RQ-7B.

The RQ-7 uses a twin-boom tail and pusher propeller configuration, carrying a variety of mission packages that can include radar, electro-optical and thermal sensors, communications relay equipment and hyperspectral sensors. It is capable of automatically tracking a target, and can carry a laser designator.

The RQ-7B model has a greater wingspan and larger tail than the original, and increased the UAV's endurance from five and a half to six or seven hours depending upon payload. It is launched using a hydraulic catapult from a rail mounted on a light vehicle, such as a High Mobility Multipurpose Military Vehicle (HMMMV). It can land on almost any flat

Above: An RQ-7 Shadow system consists of three UAVs plus a fourth disassembled as a spare. A unit consists of two light vehicles carrying the UAVs and launchers, and two more with trailers providing personnel mobility. This unit provides 72 hours of operations before needing resupply.

SPECIFICATIONS: RQ-7 SHADOW 200 TACTICAL

Length: 3.4m (11ft 2in)
Width: 3.9m (12ft 8in)
Height: 1m (3ft 3in)
Powerplant: 1 x UEL AR-741 208cc rotary piston engine

Maximum takeoff weight: 149kg (328lb)
Maximum speed: 207km/h (129mph)
Ceiling: 4570m (14,993ft)
Range: 78km (48 miles)
Endurance: 7 hours

surface providing about 100m (328ft) is available to stop in. If space is minimal, an arrester hook can be used to bring the drone to a halt more quickly.

Most operations are autonomous, although the operator can assume control over the drone at any point in the mission. The ground control station is designed for mobility and uses proven technology that was developed and adapted for the RQ-7 rather than undertaking a more risky project to create dedicated control equipment.

It is possible that RQ-7 drones have been used or will be used as part of manned/unmanned teams with Apache or other attack helicopters at some point in the near future. In the meantime, the RQ-7 was the first military drone to be issued an FAA certificate allowing it to operate at a civilian airport.

Future developments for the RQ-7 may include an armed version, probably using some small lightweight weaponry, such as the Pyros Small Tactical Munition. This is a precision air-launched bomb guided by GPS or semi-active laser homing, which would be primarily effective against fairly small personnel targets, such as a band of insurgents or a group planting a roadside IED.

ScanEagle

The ScanEagle drone is a militarized version of an earlier UAV developed to locate and track shoals of fish from the air. It is composed of separate and easily replaceable components – wings, nose and propulsion sections, fuselage and the electronics section – allowing rapid replacement of any component that becomes damaged.

ScanEagle is launched from a pneumatic catapult that can be carried on a wheeled trailer or mounted aboard a small warship. It is recovered either

Below: The original RQ-7 Shadow entered service in 2002, with the improved Shadow B following two years later. The Shadow B has longer wings that carry additional fuel, increasing endurance to around 6 hours. Upgraded sensors and electronics have also been introduced, including communications relay equipment.

SPECIFICATIONS: SCANEAGLE

Length: 1.55–1.71m (5ft 1in–5ft 6in)
Width: 3.11m (10ft 2in)
Powerplant: 1 x 2-stroke 3W piston engine
Maximum takeoff weight: 22kg (48.5lb)
Maximum speed: 111km/h (92mph)
Ceiling: 5950m (19,500ft)
Endurance: 24 hours

Left: ScanEagle's control system uses a simple point-and-click interface which can instruct the UAV to automatically track a target once the operator has indicated it. The stabilised sensor turret eliminates vibration and movement due to the UAV's flight characteristics, and can carry a range of instruments.

by belly-landing on any flat surface, or by flying it into a catcher device that snags hooks located on the wingtips. Despite requiring great precision from the differential GPS guidance unit, this system has performed acceptably on hundreds of recoveries aboard US Navy vessels.

Propulsion is by means of a rear-mounted propeller; a heavy-fuel engine is available for the improved ScanEagle2, whereas the original version is powered by standard automobile fuel. Both versions have comparable endurance – 24 hours or so – whereas the heavy fuel engine produces more electrical power for systems carried aboard the drone and is safer to store. Initially the heavy fuel engine suffered from reduced operational endurance, necessitating the development of new ignition technology.

Payload is carried in a directional turret that mounts a camera and thermal imager. In addition, a small synthetic aperture radar system was developed for use with ScanEagle. Other payloads can be quickly swapped in, including a chemical/biological detector, laser designator and a magnetic anomaly detector.

In addition to the usual applications as a tactical reconnaissance platform, ScanEagle has been trialled in the sniper-location role, operating in conjunction

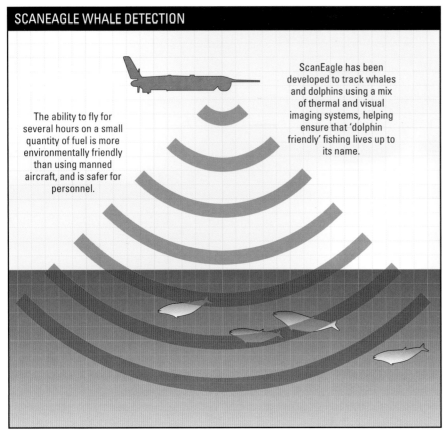

SCANEAGLE WHALE DETECTION

The ability to fly for several hours on a small quantity of fuel is more environmentally friendly than using manned aircraft, and is safer for personnel.

ScanEagle has been developed to track whales and dolphins using a mix of thermal and visual imaging systems, helping ensure that 'dolphin friendly' fishing lives up to its name.

with other instrumentation to locate the source of gunfire. This is an important role during peacekeeping operations, as well as when protecting a base or convoy within what is not always friendly territory.

ScanEagle drones have seen service worldwide with the armed forces of the UK, USA, Australia and others. It has also assisted in operations against pirates, hijackers and drug smugglers. Other applications include operations off Alaska, observing sea and ice conditions and obtaining data on the numbers and movements of whales. This is something of a return to its roots for a drone that started out as a fish-watching platform.

Above: Despite its civilian and nautical origins, the ScanEagle drone has seen extensive service on land with military forces. Hooks at the end of the wings are used to catch a recovery device and, coupled with catapult launch, this makes the UAV suitable for use aboard a small vessel or in very cluttered terrain.

Rotorcraft drones

Rotorcraft offer advantages in terms of precision and the ability to hover, which are sometimes offset by their slower speed and need to expend considerable power just to stay in the air. Rotorcraft tend to have a much lower operating altitude than fixed-wing designs; air that is too thin to allow a rotorcraft to support itself can sometimes still provide enough lift for a winged aircraft. In general, it is cheaper and simpler to build a fixed-wing UAV than a rotorcraft with equivalent performance, and in most roles a conventional craft offers advantages in terms of payload and endurance.

Above: Tiltrotor craft offer the advantages of conventional aircraft in terms of speed and economical operation in level flight, and, in addition, can hover or land vertically. The Eagle Eye UAV attracted interest from the US Marine Corps and the US Coast Guard, both of whom saw real advantages in operating this kind of aircraft.

The big advantage of rotorcraft is that they have a zero-distance takeoff and landing capability. Just as helicopters can be operated from ships that cannot support a fixed-wing aircraft, so rotorcraft drones can be operated in very tight spaces.

The majority of military rotorcraft drones are fairly large and resemble conventional helicopters in that they have a central set of rotors. These may be complex, such as a pair of separate contra-rotating rotors on the same shaft or intermeshing rotors in the style of a syncrocopter, but the overall design tends to be similar to a conventional helicopter. Smaller drones make use of multiple sets of rotors, each with their own power source, but this is ineffective for larger UAVs that need a powerful engine to lift a very significant weight.

Rotorcraft drones can, in theory, do anything that a manned helicopter can, including carrying cargo, acting as a sensor or weapons platform, or evacuating casualties. The latter is an intriguing concept, not only for the military but for disaster-response situations. It may be that in the near future casualties or survivors can be rapidly removed from danger by an automated rescue drone.

One application for this is mountain rescue and similar casualty-recovery situations. Helicopters are often used to bring casualties out once they have been located by rescuers. There is no real reason why drones could not fulfil the same function. It is not hard to imagine a 'dial-a-drone' subscription service that entitles the subscriber (for a pretty large fee, most likely) to request evacuation from whatever is going on around him using his phone.

For military forces engaged in disaster relief and similar humanitarian efforts, a stream of casualty-removal and resupply drones could link a shore party to the extensive medical and logistics capabilities of a naval force offshore. This would reduce the strain on pilots having to fly repeated missions and make numerous landings, possibly in bad weather or poor visibility.

MQ-8 Fire Scout

Helicopters have proven invaluable aboard warships since they were introduced. Capable of taking off and landing from a small space, they can be accommodated aboard a modestly sized vessel that could not carry a flight deck suitable for fixed-wing operations. While no helicopter can outperform a fixed-wing aircraft in terms of warload, speed or range, helos fulfil a number of important roles – some of which cannot be carried out by a conventional aircraft.

A naval helicopter can act as a remote sensor platform, enabling a warship to reconnoitre suspect vessels or guide

Below: The MQ-8A Fire Scout was based on an existing (manned) helicopter, significantly reducing development costs. It made history as the first autonomous landing on a moving ship by an unmanned aircraft. Previous landings had been made by unmanned craft under remote control.

missiles to a target without itself being attacked or, quite possibly, detected at all. This does expose the helicopter and its crew to danger, but the tradeoff is increased safety for the parent vessel and its crew. Of course, if the helicopter is a remotely piloted vehicle then such missions can be carried out without risking any human lives.

The MQ-8 was derived from a manned helicopter, creating a drone that takes up as much space as a full-sized helicopter but which can carry an equivalent load. Since there are no humans aboard, this means more space and lifting capacity available for payload. The original version (MQ-8A) first flew autonomously in 2000. A larger and improved MQ-8B was developed after the intended user, the US Navy, expressed doubts about the UAV's ability to meet its needs.

Interest from the US Army kept development going until the navy decided to undertake evaluation of the improved variant. In 2009, the Fire Scout went into production for use by the US Navy, although the army dropped the project the next year. The navy version was upgraded with maritime surveillance radar in 2012.

The Fire Scout UAV is a multi-role platform capable of carrying out a range

of missions that include reconnaissance, search and combat operations. Its modular payload system allows it to be quickly swapped between configurations and to receive new systems as they become available. Among these is the AN/DVS-1 Coastal Battlefield Reconnaissance and Analysis system, which is designed to search for mines and underwater obstructions in shallow coastal waters.

The Fire Scout UAV carries thermal and optical cameras in a multi-directional ball turret, along with a laser rangefinder with additional sensor packages, including maritime surveillance radar, signals intelligence and electronic warfare packages and a minefield-detection system for use against land mines.

As a weapons platform, the MQ-8B can carry Hellfire missiles and GBU-44 Viper Strike guided bombs. Its stub wings can also mount pods containing laser-guided 70mm (2.75in) Hydra rockets. These give the Fire Scout a short-range precision strike capability against small fast-moving targets, such as boats.

Armed small boats are a very significant threat in the modern battlespace. They might be used for piracy against unarmed commercial vessels, but, in some cases, have been used to attack warships. The ability to intercept and destroy such vessels before they can reach weapons range, i.e. to 'break the kill chain', greatly simplifies a naval commander's force protection problems.

The world's navies increasingly operate in littoral waters – close to the shore – and face a variety of threats. The US Navy plans to use Fire Scout UAVs for force protection missions from various platforms, including the Littoral Combat Ship, either operating alone or teamed up with manned helicopters. Rapid identification of targets as hostile or otherwise, and short response times if a threat is detected, are vital when operating in this cluttered environment, which may contain many civilian vessels. A pair of MQ-8 Fire Scouts operating in relay can constantly monitor an area up to 110 nautical miles (204km/126 miles) from the parent vessel.

The MQ-8 Fire Scout was the first unmanned air vehicle to land autonomously aboard a US Navy warship. No pilot interaction is required during landing and takeoff operations, which can be carried out while the parent vessel is manoeuvring. This capability allows the UAV to act as an unmanned supply aircraft, a role envisaged for the enlarged MQ-8A version. Development of this capability was not without incident; in 2010 a Fire Scout ceased responding to commands and entered restricted airspace over Washington DC. It was not carrying weapons, but, all the same the incident rekindled a certain amount of debate about the use of unmanned vehicles in populated areas.

Fire Scout UAVs have served in a variety of roles, acting as reconnaissance platforms in Afghanistan and during the 2011 intervention in Libya. During piracy-suppression operations off the west coast of Africa, Fire Scout drones were used for monitoring and surveillance, including near-constant observation of some areas using drones in relay. Other MQ-8 operators have successfully detected and identified fast boats engaged in drug smuggling, leading to interception.

The MQ-8 Fire Scout is perhaps unusual in that it is not a single design – A, B and C variants were built around different airframes. The fact that it is not a custom-designed UAV, but a manned aircraft converted to autonomous operations, is also intriguing. This approach has typically not been used by UAV developers, but it may be that other conversions will appear in the near future.

A-160 Hummingbird

The Hummingbird UAV was initially developed using a converted version of a commercially available light helicopter, starting in 1998. Unusually for such projects, development went straight to unmanned testing rather than keeping a human pilot aboard to deal with any incidents that might occur. Although the original test drone was lost in a crash, results were promising enough to commission further testing with what would become the A-160 Hummingbird. Small numbers of the test drone, which was named Maverick, were taken by the US Navy for operations, but details were not made public.

The A-160 Hummingbird first flew in late 2001, and, although more test vehicles were lost, the project produced an innovative rotorcraft that used variable-speed rotors to achieve efficient flight characteristics at varying altitudes. By altering the rotors' speed of rotation the A-160 can optimize its fuel efficiency or maximize lift depending on where it is operating. In 2008, the Hummingbird UAV made a flight of over 18 hours and landed with some fuel remaining, setting a record for the longest flight made by any rotorcraft. Hummingbird also demonstrated an ability to hover at altitudes of up to 6100m (20,000ft).

The A-160 Hummingbird was evaluated for service with the US Army, US Navy and US Marine Corps as a transport drone, as well as a sensor platform. A foliage-penetrating radar was developed to find targets in cluttered terrain such as rain forests. US Special Forces have taken several examples and apparently plan to purchase more, perhaps not least because the low-speed rotors of the Hummingbird produce a

very low acoustic signature. Noise is one of the main drawbacks with using helicopters for covert operations, so a quiet drone helicopter makes a lot of sense in this environment.

Although the rotors and transmission are of innovative design, their configuration is conventional. Hummingbird uses a four-bladed main rotor and a tail rotor for steering. The fuselage is largely constructed of carbon fibre that is both light and produces a low radar return.

Payloads include thermal and electro-optical cameras, synthetic aperture radar and a laser designator. Electronic warfare and communications packages can also be carried. Most functions, including takeoff, landing and navigation, are autonomous, with all aspects of the engine and transmission automatically handled by the flight systems.

Above: The A-160 Hummingbird can adjust its rotor operation depending on the requirements of the situation. This has been compared to changing gear in a car as speed increases, and greatly enhances performance and endurance over conventional rotor systems.

SPECIFICATIONS: A-160 HUMMINGBIRD

Length: 10.7m (35ft)
Main rotor diameter: 11m (36ft)
Powerplant: 1 x Pratt and Whitney Canada PW207D
Maximum takeoff weight: 2948kg (6499lb)

Maximum speed: 258 km/h (160mph)
Ceiling: 6100–9150m (20,000m–30,000ft)
Maximum range: 2,589km (1609 miles)
Endurance: 18+ hours

Above: The APID-55 UAV began development in the early 1990s, and first flew in 2008. It is designed to operate in a range of conditions from desert to arctic and, like other small rotorcraft drones, requires very little space to take off or land.

COMPARISON OF MAXIMUM TAKEOFF WEIGHTS OF ROTORCRAFT DRONES

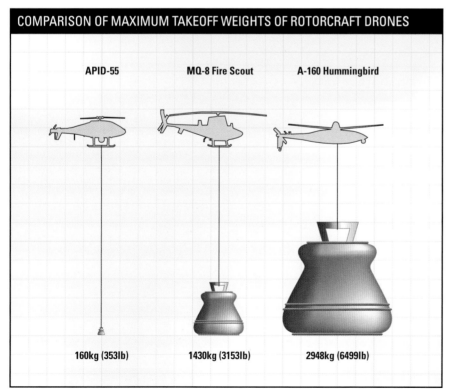

APID-55	MQ-8 Fire Scout	A-160 Hummingbird
160kg (353lb)	1430kg (3153lb)	2948kg (6499lb)

APID-55 Family

The APID-55 small rotorcraft UAV began development in the 1990s, and first flew in 2008. It was developed with military and non-military applications in mind, including monitoring of emergencies such as fires, search and rescue and scientific/environmental observations. In a military context, it is designed for reconnaissance and patrol/border monitoring work, and can conduct mine detection operations.

The APID-55 is constructed of lightweight materials, such as titanium, aluminium and carbon fibre, and uses a conventional main rotor and tail rotor configuration. These are powered by an internal combustion engine with enough fuel for a six-hour flight. The larger, faster APID-60 version is distinguishable by its use of skids rather than wheels for landing gear.

There are plans to upgrade the APID-60 with a heavy fuel engine, allowing it to use the same fuel as all other vehicles in keeping with NATO logistics policy. The difficulties in creating a heavy-fuel engine suitable for small craft are not inconsiderable, mainly due to power-to-weight issues with a small engine.

The APID-55 and APID-60 carry stabilized electro-optical and thermal cameras suitable for a range of surveillance and reconnaissance missions. Navigation is mostly autonomous, using preset waypoints and GPS guidance. Waypoints can be amended in flight by the operator, and the payload can be remotely operated from the ground. APID carries infrared and barometric altimeters, in addition to its GPS unit and a laser scanner, and has been successfully tested in environments ranging from hot desert to the arctic.

The APID-55 was initially developed to meet the needs of the United Arab Emirates Defence Agency, but is proving quite versatile. Potential users range from customs and border protection agencies – the Chinese customs service is reported as buying a number of APID drones – through oil companies wishing to conduct pipeline inspections remotely, to military services.

Below: APID-60 is an improved version of APID-55, with a 20km/h (12.4mph) higher top speed. The gimballed sensor turret containing electro-optic and infrared cameras is carried under the fuselage between the landing skids. Internal instruments include a GPS receiver plus infra-red and barometric altimeters.

Transport and utility drones

It has been remarked that while amateurs are fascinated by tactics, professionals are wearily obsessed with logistics. Without resupply, any military effort will simply grind to a halt. Even when engaged only in day-to-day operations, a military force gets through an enormous quantity of food, boots, clothing, tools and maintenance spares, and all manner of sundry items that must be replaced on a constant basis. All modern military forces have a large 'tail' supporting their combat elements, or 'teeth'.

Above: Most supplies move on the ground in trucks, which are vulnerable to ambush. Significant additional resources are absorbed in escorting the convoys and securing their routes. Rapid, low-cost air supply operations using UAVs may offer a viable alternative that does not expose support personnel to risk.

Procurement and ordering are largely automated, but when it comes to getting supplies to where they are needed the traditional guy-in-a-truck method seems unbeatable. However, this absorbs a lot of manpower and exposes personnel to the risk of ambush or attack with roadside IEDs. There is also the prospect that tired truck drivers may have an accident or simply take a wrong turn in a confusing environment. This has happened in recent history, taking a convoy into insurgent-held territory with disastrous consequences.

The concept of using automated systems to resupply military personnel in the field is intriguing. GPS-guided transport drones do not fall asleep at the wheel or take wrong turns. Nor do they decide not to reenlist because they were bored to screaming point driving the same truck down the same road for months on end.

Arguably, if some of the logistics function can be taken by drones, this will free personnel to perform tasks that no machine can yet handle. If the tail can be automated, then perhaps the military machine can grow a few more teeth.

AirMule

AirMule was developed in Israel to meet a need that arose during conflict in Lebanon in the mid-2000s. Military units needed a fast and efficient means of transporting casualties out of the combat area and obtaining resupply in cluttered urban terrain where helicopters could not operate.

The AirMule UAV uses an innovative internal rotor system that enables it to operate in spaces far too small for a conventional helicopter. It is held aloft by two large downward-facing fans on the bottom of the chassis, with a pair of smaller steerable fans for propulsion and direction. There are no moving components outside the vehicle's ground footprint, which is only slightly larger than that of a HMMMV, therefore there is no hazard to nearby personnel from rotors.

Below: An AirMule UAV can deliver 500kg (1100lb) of cargo in a single mission, with an operating radius of 50km (31 miles). The designers envision a supply unit using several of these vehicles to provide constant resupply and casualty evacuation, with a theoretical capacity of 6000kg (13,200lb), per AirMule, per 24-hours.

Above: AirMule is an innovative craft that uses internal fans rather than a set of rotors that extends beyond the fuselage. This enables it to get into spaces that a helicopter could not. One potential application of this capability is casualty evacuation for both military and civilian emergency situations.

The AirMule UAV is designed for a variety of transportation tasks. For military and non-military users, these include casualty evacuation, delivery of supplies and routine personnel transfers. It can function in bad weather and can hover in wind conditions of up to 50 knots (93km/h; 57mph). Its small size and low signature are mainly advantageous to military users, although reduced noise may be a factor for civil and commercial operators.

The AirMule has an endurance of about two to four hours and can fly at up to 3660m (12,000ft). In the event of a transmission or rotor malfunction, or the failure of its single engine, AirMule has an automatically deployed parachute system designed to bring the craft down safely from any altitude up to its maximum operating height. If control signals are lost, the UAV's flight systems will land it.

Although not a reconnaissance asset, AirMule requires a great deal of information about its environment. It has two laser altimeters and radar systems for navigation and target indication, plus GPS and inertial navigation systems. Control can be performed manually from the ground control station, or the UAV can operate autonomously when required.

AirMule's designers envisage a large-scale logistics operation capable of keeping a major combat force supplied in the field and bringing out casualties or personnel who need to transfer to the rear for other reasons. This has the advantage of reducing vulnerability to mines and roadside IEDs, while reducing

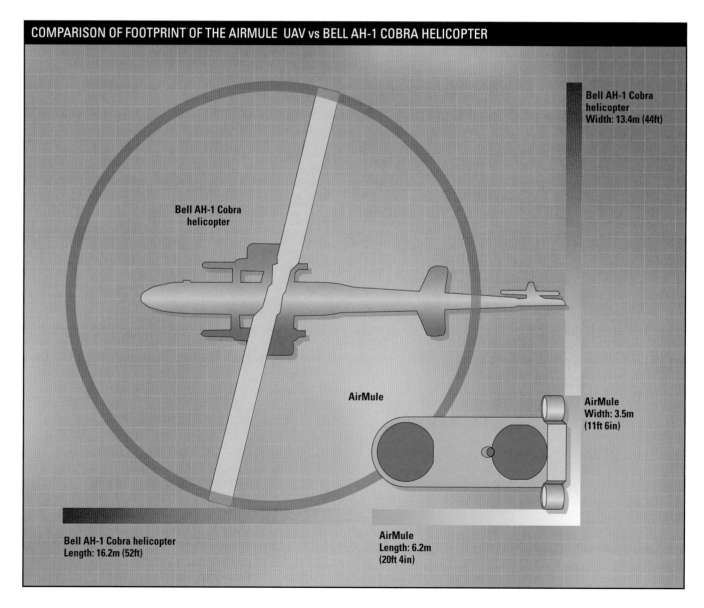

COMPARISON OF FOOTPRINT OF THE AIRMULE UAV vs BELL AH-1 COBRA HELICOPTER

Bell AH-1 Cobra helicopter Width: 13.4m (44ft)

Bell AH-1 Cobra helicopter

AirMule

AirMule Width: 3.5m (11ft 6in)

Bell AH-1 Cobra helicopter Length: 16.2m (52ft)

AirMule Length: 6.2m (20ft 4in)

Above: The paired rotors of this K-Max intermesh, but cannot collide, as they are driven from a common power source. The synchrocopter layout eliminates the need for a tail rotor and the machinery associated with it, which saves weight, and provides greater stability than a conventional helicopter.

SPECIFICATIONS: AIRMULE

Length: 6.2m (20ft 4in)
Width: 3.5m (11ft 6in)
Height: 2.3m (7ft 7in)
Main rotor diameter: 1.8m (6 x 5ft 11in)
Powerplant: 1 x Turbomeca Ariel 2 turboshaft turbine
Maximum takeoff weight: 1406kg (3100lb)
Maximum speed: 180km/h (112mph)
Ceiling: 3660m (12,000ft)
Endurance: 2–4 hours

pilot or driver fatigue, and enables high-risk missions to be undertaken without endangering the lives of a human crew.

K-Max

The K-Max is an Optionally Piloted Vehicle (OPV) that, as the name suggests, can operate autonomously or with a human pilot aboard. It is developed from a well-proven rotorcraft design that was designed from the outset as an 'aerial truck'. Some of its features, such as a cockpit designed to allow the pilot to see an underslung load, are not relevant to autonomous operations, but neither do they detract from it.

The K-Max is a syncrocopter. It has two sets of rotor blades that intermesh, but cannot collide, rotating in opposite directions on the same axis. This eliminates one of the key problems

encountered by conventional rotorcraft – the torque created by the main rotor. This can cause the aircraft to rotate in place and must be counterbalanced by a tail rotor. A syncrocopter does not need one, nor any of the power transfer mechanisms associated with it.

Synchrocoptors deliver powerful lifting capacity for their engine size, and are extremely stable when hovering. This has made them a popular choice for transport work in industries, such as logging, where precise positioning of a load suspended under the aircraft can be vital.

Above: Applications of unmanned helicopters go beyond cargo transportation. A UAV can deliver a sensor package or a communications relay unit to a remote area, or even an unmanned ground vehicle. For missions requiring a hands-on approach, K-Max can be flown in the normal manner by a pilot.

The autonomous (or optionally piloted) version of the K-Max synchrocopter was deployed in Afghanistan on a trial basis as a 'flying truck'. While unable to carry large loads as cost-effectively as ground trucks, K-Max drones are effectively immune to roadside IED attacks. This is a major consideration for military units that have to be regularly resupplied; something as simple as delivering rations is a hazardous trip for truck crews.

K-Max OPVs flew routine resupply missions in Afghanistan from 2011 until 2014, averaging about five a day at one point. Many flights took place at night and did not encounter enemy action, although a K-Max did crash due to a combination of weather conditions and a swinging heavy load. Despite this accident, the programme was generally considered a success.

At the time of writing it is unclear whether either the US Army or US Marine Corps will proceed with procurement of autonomous delivery vehicles. The concept seems workable and promising,

Above: The K-Max UAV is based on a helicopter designed to carry heavy loads on a sling. This creates some challenges for the pilot or control system – a swinging load in a high wind is one of the most difficult scenarios such a craft can encounter.

and certainly the K-Max proved invulnerable to the roadside IEDs that became a hallmark of the conflict in Afghanistan. For the US Marine Corps in particular – or any force conducting amphibious operations – the ability to deliver supplies from offshore landing ships and logistics vessels is, of course, vital. Doing so autonomously may offer advantages in terms of reducing pilot fatigue and risk to personnel. Likewise, automatic resupply of otherwise isolated outposts and bases might be an effective way to prevent needless casualties during necessary but routine supply runs.

However, at the present time there is no clear and definite need for an autonomous or optionally piloted helicopter. Various nations have expressed an interest in the concept, and work is ongoing to increase capabilities. Enhancements may include automated threat-avoidance systems and the ability to retask delivery aircraft in flight. Operations integrating numbers of autonomous delivery drones, either

in formation or relays, may prove highly beneficial, and would be enhanced by an automated hook-up system for payload that is under development.

Operations in Afghanistan proved conclusively that the unmanned delivery vehicle can and does work in a difficult operating environment. The next question is whether there exists a perceived need to fill this niche, and whether it can be done in a cost-effective manner.

Left: The designers of Camcopter S-100 are aiming at more than one market niche. In addition to military applications, UAVs have national security applications, such as preventing smuggling or illegal border crossings, and can also be used to monitor the spread of contamination in the event of an oil spill.

Camcopter S-100

The Austrian-designed Camcopter S-100 was developed to meet the needs of the German Navy and the military of the United Arab Emirates for a multi-role medium-altitude, medium-range rotorcraft UAV. Other nations have also adopted the Campcopter S-100, including China, Italy and Russia.

Camcopter S-100 is designed to operate from maritime vessels using a ground control station that has two systems: one that handles mission planning and UAV control, while the other deals with data retrieval and payload control. The mission planning system can keep track of hazards and no-fly areas; hazard areas (e.g. zones covered by anti-aircraft weapons) can also be marked for avoidance.

Camcopter can fly autonomously or under direct control using a joystick on the control unit. It also has a vertical takeoff/landing system and an automated return-to-base mode. It is capable of operations in a radius of about 180km (112 miles) from base and lasting up to six hours.

In addition to electro-optical and thermal sensors, Camcopter carries synthetic aperture radar and a LIDAR (Laser Imaging Radar) system. It can also carry ground-penetrating radar, which can be used to search for mines or buried IEDs. Payload can be carried in two payload bays or side-mounted hardpoints, and there is an auxiliary avionics bay for additional flight electronics. Other upgrades include an enlarged fuel capacity and the ability to carry an under-slung load, allowing the

Camcopter UAV to function as a delivery drone. A heavy-fuel engine is also available, bringing the Camcopter into line with NATO logistics procedures and also increasing the safety of fuel storage. This is considered particularly important for users wishing to deploy the UAV aboard ships.

Camcopter S-100 is able to fulfil a variety of possible roles, depending on the end user's needs. Thus far it has primarily found favour with naval forces, becoming the first UAV to fly from an Italian warship. The ability to detect mines and IEDs makes it attractive to land forces operating in environments such as Afghanistan and Iraq where these weapons are in use by insurgents.

It has also been suggested that UAVs, such as Camcopter, could take a more assertive role in law enforcement than acting as airborne camera platforms, perhaps, for example, delivering tear gas to disperse a riot. This would have the advantage of not exposing law enforcement personnel to risk, but might be a public-relations issue. The image of black-painted drones resembling flying sharks, dropping tear gas on crowds below, might become adversely associated with governmental oppression. Alternatively, it might be seen as reassuring and extremely impressive. Much depends on point of view, but what is certain is that it will be controversial and sure to provoke a certain amount of heated debate.

SPECIFICATIONS: CAMCOPTER S-100

Length: 3.1m (10ft 2in)
Height: 1.12m (3ft 6in)
Main rotor diameter: 3.4m (11ft 2in)
Powerplant: Rotary engine
Maximum takeoff weight: 200kg (440lb)
Maximum speed: 240km/h (150mph)
Ceiling: 5496m (18,000ft)
Endurance: 6 hours with 34kg (75lb) payload, plus optional external fuel tank to extend endurance to 10 hours

Small reconnaissance drones

Small reconnaissance drones are designed to provide tactical reconnaissance using a fairly basic array of sensors. There are, of course, severe limitations on what can be achieved with such a small vehicle, but this is offset by a low cost that makes them affordable in large numbers. The benefits of a mobile airborne camera are considerable, especially during counterinsurgency operations carried out in cluttered terrain that blocks line-of-sight visibility.

Above: Whilst the capabilities of small UAVs, such as Desert Hawk, are limited compared to the large, long-range drones, they are inexpensive and easy to operate. For a modest investment, small UAVs provide ground forces with a measure of airborne reconnaissance capability that might not be otherwise available.

Whereas an expensive high-capability reconnaissance UAV can obtain strategically important information and an armed drone can act upon it, a small reconnaissance UAV is more of a 'force-multiplier' for local formations on the ground. Small drones of this sort can warn of potential intruders near a base or an ambush just ahead, or it can guide a ground patrol towards an elusive enemy. They can also be used to assess damage after a strike by aircraft or artillery, or to assist short-term mission planning by providing timely intelligence about what is going on nearby.

Thus, while the capabilities of these small and sometimes funny-looking drones are very limited, their effects are considerable if well handled. They have relatively limited value on their own, but when integrated into ground operations they can increase effectiveness and efficiency, and sometimes prevent a disaster by providing a warning that would otherwise not be available.

Desert Hawk

Resembling a rather tubby toy plane, the Desert Hawk UAV was designed with

Above: The amount of equipment a ground force can carry is of course limited, but a Desert Hawk UAV and its control station take up little space and weight. The increase in effectiveness for ground forces thus equipped is often worth losing whatever other equipment might have been carried.

security and short-range reconnaissance operations in mind. It uses an electrically driven pusher propeller and is launched by hand with the assistance of a bungee cord. After Desert Hawk's mission, which has a maximum duration of about one hour, it lands on its belly on whatever surface is available.

The Desert Hawk UAV does not have wheels, but instead uses Kevlar skids. It is light and tough, being composed mainly of expanded polypropylene, and will handle a fairly rough touchdown. Landing speed is low, enabling the operator to bring the drone down in

a fairly small space without needing a runway. Desert Hawk is also surprisingly tolerant of bad flying conditions, which is necessary when operating in some environments where high winds and turbulent air are common.

Desert Hawk carries a sensor package, including infrared and electro-optical systems, as well as a laser illuminator that allows video images to be taken in total

darkness using night vision equipment. It has a plug-and-play modular architecture for additional equipment and can carry SIGINT and COMINT packages or a synthetic array radar.

Control is by way of a laptop-type interface, which receives imagery from the craft as well as transmitting control signals. Most flight operations are automatic, using GPS navigation, with

Right: The UAV component of an RQ-11 Raven system accounts for about 15 per cent of its total cost. The control system and ground antennae are far more expensive, but fortunately are the components least likely to require replacement. A UAV operating in enemy territory must be considered at least potentially expendable.

The Desert Hawk UAV was particularly useful in the base-protection role as it proved to be capable of identifying personnel carrying shoulder-fired missiles up to 10km (6 miles) away. This greatly reduced the possibility of insurgents taking up positions close to an airbase to launch a 'Stinger ambush' on aircraft taking off or landing. Similarly, ambushes on armoured forces or supply convoys can be detected much earlier and avoided or counterattacked if the presence of weapons can be determined from a distance.

RQ-11 Raven

The RQ-11 Raven first appeared in 1999 under a different designation, and later matured into its present form. It was adopted by the US Army for short-range tactical reconnaissance in 2005, and since that time it has been taken into service by many international operators as well as other branches of the US military.

The Raven has become more or less ubiquitous, and is now serving with many countries worldwide. It is a fairly simple design resembling a model aircraft with a high wing and a pusher propeller. This is driven by a small electric motor that has enough battery power for flights of 60–90 minutes' duration.

Payloads include forward and side-looking electro-optical or thermal cameras, which transmit data back to the control station. The UAV can navigate autonomously, or be directly controlled by the operator out to a distance of about 10km (6 miles). It has an automatic landing system and, in an

Above: Uninformed observers might be forgiven for considering the rather unglamorous nature of small UAV operations to be nothing more than messing about with expensive toy planes, but these are serious pieces of military kit that can offer very real advantages to their users.

the operator inputting commands as needed. First flying in 2003, Desert Hawk was adopted by the British Army for use with artillery formations and by the US Air Force for base protection. The latter role is very manpower-intensive, and the need to

provide physical patrols can be debilitating over time, especially in very hot conditions. Wherever possible, cameras and other electronic means are used to reduce manpower requirements while maintaining or enhancing security levels.

Above: Over 3000 RQ-11A Raven UAVs were produced up to 2006, when the upgraded RQ-11B version became available. Raven can land more or less anywhere by stalling close to the ground and dropping a short distance. It is light enough not to be damaged, which also permits it to be launched by hand.

SPECIFICATIONS: RQ-11 RAVEN

Length: 91.5cm (3ft)
Wingspan: 1.37m (4.5ft)
Powerplant: Aveox 27/26/7-AV electric motor
Weight: 1.9kg (4.2lb)
Maximum speed: 48–96km/h (30–60mph)
Range: 10km (6 miles)
Ceiling: 30.5–152.4m (100–500ft)
Endurance: 60–90 minutes
Launch method: Hand-launched

emergency, can be ordered to land itself with a single command.

Upgraded versions of Raven have appeared over time, including Raven Gimbal, which uses a small gimballed sensor turret to carry both IR and visual cameras. The user can swap between data feeds instantly while the UAV autonomously adjusts its flight path to accommodate the choice of target. Other projects include the addition of solar panels to the upper wing surface. Using these to power the onboard electronics reduces battery drain and increases endurance.

In US Air Force service, Raven drones are deployed with two-man operator teams who carry a pair of UAVs and their control equipment in backpacks. The

drone can be quickly readied for use and is hand-launched, i.e. it is simply thrown by the operator. It flies relatively low – no higher than 150m (500ft) – and is thus vulnerable to small arms fire, if spotted. However, drones of this sort have proven extremely useful and are affordable in sufficient numbers to offset any losses suffered in action.

Avinc Puma

The Puma UAV is another 'toy plane' type drone, this time using a two-bladed tractor propeller. Unlike a civilian model plane, however, it is designed for the adverse conditions that might be encountered in a military environment. Where a model plane enthusiast might decide not to fly on a given day due to bad weather, military forces need reconnaissance data at all times and therefore require a drone that is capable of operating in difficult conditions.

Puma is hand-launched, after which it uses a fuel cell system for propulsion that can be recharged between missions. The airframe and other systems are designed to be tough enough to survive attempts to fly the drone in bad weather. This also makes the UAV relatively 'soldier-proof'. It can land on its belly on level ground or in water.

Endurance depends upon speed and conditions, but in tests the Puma

Above: The 'AE' in RQ-20A Puma AE stands for 'all environment'. This small UAV can land on water or any level ground, and launch requires only enough space to take a few steps and fling the drone into the air. Its construction is robust enough to survive operations in a military environment.

Below: Puma can carry a payload in a 'transit bay' under the wing. This is in addition to its internal sensor package, consisting of gimballed thermal and electro-optical cameras, and permits new capabilities, such as, communications relay to be quickly added or removed as the mission profile requires.

UAV has achieved flights of five to nine hours' duration with fuel cells aboard and around two hours using a rechargeable battery. Its waterproof electronics package includes infrared and electro-optical cameras, and a GPS guidance unit for navigation. Data is passed in real time to the ground control station, out to a maximum operating radius of about 15km (9 miles). The ground station is capable of pulling still images from video and retransmitting data to other users. Puma drones have been taken into service by U.S. Special Operations Command (SOCOM). It is used for target identification, tactical reconnaissance and damage assessment, but has wider applications, including anti-smuggling and border patrol missions, maritime surveillance and search and rescue.

Wasp

Wasp is a very small UAV developed for the US military primarily for route reconnaissance, tactical reconnaissance and force protection missions. Force protection is a constant concern for all military units, although it is never the primary mission. That is to say, no effective military force ever sallied forth with the primary mission of getting home safely. Were that the main objective, staying at home would probably be a better option.

Military forces always have a primary mission that might be anything from striking at an enemy to rescuing survivors after a disaster, but must also be mindful of the need to remain as safe as possible and to avoid pointless casualties. 'force protection' is all about minimizing risk and enabling a rapid response to any situation that develops. Tactical information is a vital tool in this endeavour.

The Wasp UAV supports the force protection mission by providing tactical reconnaissance at the squad level. To do this, a UAV has to be small enough to be carried by a squad member without reducing the amount of other equipment he can carry or his ability to move quickly and use his weapon. Thus a drone intended for this role must be very small, which precludes a heavy payload or a long endurance. It must also be inexpensive and capable of being deployed quickly.

Wasp is land-launched and can remain airborne for about 45–90 minutes and has an operating radius of around 5km (3 miles). It navigates autonomously and can land without user guidance. A water-landing-capable version is currently under development. Payload consists of visual and thermal cameras that provide real-time data to the control unit. This is the lightweight Common Ground Control Station also used with the Puma and other UAVs. If contact with the ground station is lost – for example, due to the cluttered urban terrain in which many operations take place – the UAV is programmed to make a safe landing autonomously.

Dragon Eye

Dragon Eye is a twin-engined small UAV driven by two electrically powered 'tractor' propellers on its wide wings. It has no tailplane, relying on the shape and size of its large wing instead. The central fuselage is wide and blocky to maximize equipment space.

Dragon Eye was created to meet

SPECIFICATIONS: PUMA AE

Length: 1.4m (4ft 6in)
Wingspan: 2.8m (9.2ft)
Powerplant: Battery
Weight: 6.1kg (13.5lb)
Maximum speed: 37–83 km/h (23–52mph)
Range: 15km (9.3 miles)
Ceiling: 152m (500ft)
Endurance: 3.5+ hours
Launch method: Hand-launched, rail launch (optional)

Above: The Wasp MAV (Micro UAV) shares a common control system with the Raven, Puma and Swift UAVs. Despite its small size, it can carry a useful payload of thermal and electro-optical cameras, and has the capability to quickly swap payload systems if the need arises.

COMPARISON OF ENDURANCE OF SMALL RECONNAISSANCE DRONES

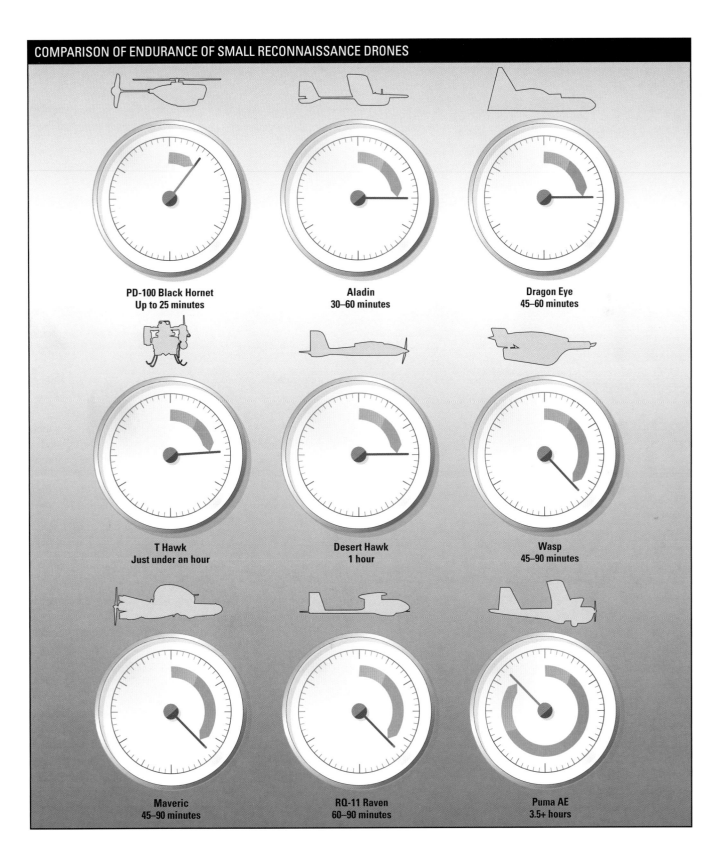

PD-100 Black Hornet
Up to 25 minutes

Aladin
30–60 minutes

Dragon Eye
45–60 minutes

T Hawk
Just under an hour

Desert Hawk
1 hour

Wasp
45–90 minutes

Maveric
45–90 minutes

RQ-11 Raven
60–90 minutes

Puma AE
3.5+ hours

Above: The Dragon Eye UAV is designed to come apart if it strikes a hard object, hopefully with the components remaining intact to be reassembled later. If one or more components do break, swapping in a replacement is easy due to this modular construction.

the needs of the US Marine Corps for a lightweight, small, tactical reconnaissance drone primarily for use in MOUT (Military Operations in Urban Terrain) situations. It is carried in a backpack and can be readied for flight in about 10 minutes. Launch is by hand or using a bungee cord, after which the drone carries out a pre-planned mission using GPS guidance and a series of waypoints set by the operator.

Dragon Eye was designed to be

as simple to operate as possible, with user training requiring less than a week. It employs a sideways-looking low-light camera to undertake tactical reconnaissance, feeding live video back to the operator over an operational radius of about 10km (6 miles).

Dragon Eye was designed to be robust, using tough and lightweight materials. Its construction is also designed for survivability – the drone comes apart rather than suffering

breakage of its component parts. A standard Dragon Eye package includes two UAVs and a ground station, with two replacement nose units to allow the most likely to be damaged component to be swapped out.

There are plans to create an upgraded version with different sensors and an automatic landing system, along with improved power units that will increase endurance beyond the current 45–60 minutes. In the meantime, Dragon Eye has found additional applications beyond the military. In 2013, it was used to study volcanic plumes as part of a project intended to improve the safety of people living or working close to a volcano.

One serious hazard in this environment is 'vog' – volcanic fog – that contains significant amounts of sulphur dioxide. The use of manned aircraft to study such conditions is very hazardous, partly due to effects on the crew and partly because aircraft engines can be choked by the air conditions. An electrically powered drone suffers from neither problem; Dragon Eye was able to collect data from areas that were simply too hazardous to send in a manned aircraft or a team on the ground.

Aladin

The Aladin UAV was developed for use by the German Army for short-range tactical reconnaissance. Using a simple wide-wing design driven by a 'tractor' propeller, the drone can be transported in two cases and readied for flight in five minutes. It is launched by hand or with the aid of a bungee cord and, has an operational radius of abut 15km (9 miles).

Once launched the UAV flies a preset

Below: Once launched using a bungee cord, the Dragon Eye UAV navigates its way through a series of waypoints set by the operator. These can be reset whilst the drone is in flight, enabling the operator to guide Dragon Eye to any point of interest, or to bring it home early.

Above: Aladin is deployed as a package containing two UAVs and one ground control station. It has been in service with the German Army since about 2005 and the Dutch Army since 2006. Both services have used Aladin in Afghanistan, for short-range aerial reconnaissance and surveillance.

course via GPS waypoints unless the operator chooses to update the navigation points. Payload includes a thermal imager and multiple cameras providing downward and sideways views, as well as forwards. Mission endurance is limited by battery life and is in the range of 30–60 minutes, which is entirely sufficient for tactical reconnaissance at the small-unit level.

An innovative submarine-launched version has also been developed. This consists of three Aladin drones and a catapult launcher mounted on a retractable mast. A submerged submarine can extend the launcher

above the surface and launch a UAV to conduct local reconnaissance. There is no real prospect of recovering the drone, which must be considered expendable.

The launching submarine could stay at shallow depth with its communications mast extended and guide the UAV while receiving real-time data, or could run deeper for a time and come up again to receive a download of the local situation. The use of a UAV might enable a submarine to conduct reconnaissance of conditions inland, or to obtain a view of the surroundings from a higher vantage point than the sail of a surfaced submarine provides.

Perhaps the most useful role for this drone is for Special Forces' teams deployed by submarine. A UAV could provide up-to-the-minute reconnaissance data for the target area, allowing the team to deploy in safety and be forewarned of any hazards. Similarly, it might be used to watch for the return of a team operating ashore and assist in guiding them safely through the exfiltration process.

Maveric

The Maveric drone uses a wide, high-wing configuration and rear-mounted pusher propeller powered by an electric motor. While this is fairly conventional for small reconnaissance drones, Maveric does have one unusual feature. Described by its manufacturers as 'biologically camouflaged', the UAV is intended to be mistaken for a bird.

The main threat of detection for small drones is being visually spotted; such a small object flying close to the ground is unlikely to be detected by radar, and thermal signature is minimal. Even if there is relatively little chance of hostiles bringing down the drone with small arms fire, detection means that the targets know they are being observed and that the drone operator is in the vicinity. At very close ranges, the Maveric will not fool an observer – few birds are equipped with a propeller – but at even a modest distance a flying object that is shaped like a bird and generally behaves like one may completely escape notice.

Maveric is hand-launched and recovered using a net or by belly-

Above: Aladin was designed from the outset to undertake day and night operations in adverse conditions. Particular requirements were the ability to cope with conditions encountered in the mountains of Afghanistan and to be capable of a quick turn-around. Aladin can be set up in a few minutes and readied for another mission by swapping its batteries.

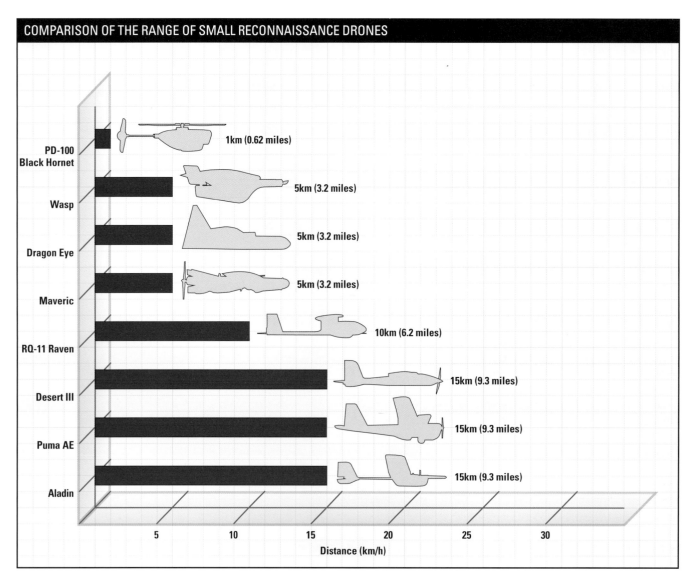

COMPARISON OF THE RANGE OF SMALL RECONNAISSANCE DRONES

PD-100 Black Hornet — 1km (0.62 miles)

Wasp — 5km (3.2 miles)

Dragon Eye — 5km (3.2 miles)

Maveric — 5km (3.2 miles)

RQ-11 Raven — 10km (6.2 miles)

Desert III — 15km (9.3 miles)

Puma AE — 15km (9.3 miles)

Aladin — 15km (9.3 miles)

Distance (km/h)

landing on the ground. Its endurance is roughly 45–90 minutes of flight, although changing the batteries takes less than a minute and relaunch can be performed from wherever the operator happens to be standing. The UAV is constructed from tough, lightweight materials that allow it to be bent without damage. This not only protects it in the event of a crash or collision, but also allows the drone to be carried in a tubular package and deployed in less than two minutes.

The standard Maveric drone mounts a camera facing forward and a second sensor in a fuselage pod. This can carry either another camera or a thermal sensor in a sideways-looking mounting, or a retractable gimballed mount containing the additional sensor. The fuselage, tail and nose sections and payload bay are all modular, with carbon fibre used for most structures.

The forward-looking camera is used for automatic collision detection and avoidance, which is controlled automatically from the Merlin ground control unit. This receives images and processes them to reduce the effects of

camera shake, and transmits operator commands out to a distance of about 5km (3 miles). Maveric can fly beyond command range, storing imagery onboard until contact is regained with the control unit.

Control is simple, using a hand-held unit that allows manual control or a more automated operation using waypoints for the selection of flight modes. These include a simple flight mode where the drone automatically maintains its height and speed, with the operator indicating where he wants the drone to go next.

Maveric can also be set to 'loiter', circling a waypoint, or to 'rally' to a marked point. It also has a 'home' mode that instructs the drone to fly to a previously determined point.

Maveric can carry out the usual tasks of a small recon drone – tactical reconnaissance, situational awareness, battle damage assessment and so forth – with the added advantage that the enemy might not recognize it for what it is. Insurgents are more likely to be careless about displaying weapons, or may choose not to conceal themselves (or their activities) when they think they are not being observed.

This ability to undertake covert surveillance may also assist when setting up an ambush or raid, as hostiles might otherwise be tipped off by the presence of a short-ranged reconnaissance drone that troops are in the area and something may be about to happen. The use of recon drones is extremely valuable, but it can be something of a double-edged sword if they are spotted and recognized. Most small drones are hard to spot; Maveric augments this by pretending to be a bird.

The T-Hawk (Tarantula Hawk)

The T-Hawk (Tarantula Hawk) drone is unusual in that it uses ducted fans rather than propellers to keep it aloft. Its hover-and-stare capability enables the user to undertake a close examination of an area without touching the ground. This is highly useful in Explosive Ordnance Disposal (EOD) operations or when searching for suspected IEDs.

Small enough to be carried in a backpack, the T-Hawk drone has an endurance of a little less than one hour, but can fly fast enough to cover a wide area in that time. In addition to its military applications, it is suitable for various security roles, and has also been used in civilian disaster-management.

In 2011, T-Hawk drones were used to assess damage to the Fukushima nuclear plant. The UAVs were able

Above: Maveric is carried in a tube with the wings folded, and extends them as it is pulled out. Once thrown into the air, it follows its programmed course and avoids obstacles autonomously using its forward-looking camera. If a collision does occur, the UAV's design is very resistant to damage.

Below: Maveric has been described as being 'biologically camouflaged', which translates as meaning it looks a lot like a bird. With virtually no noise to give it away, the UAV may not be noticed at all but if it is, observers might well discount it, since it looks nothing like a piece of military hardware.

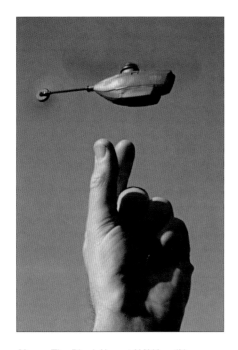

Above: The Black Hornet UAV is a 'Nano Unmanned Air Vehicle', i.e. it is even more tiny than the small drones referred to as Micro Air Vehicles. Small size limits the weight that can be carried and the drone's endurance, but it can still carry a camera in missions lasting up to 25 minutes.

to operate in the interior of the plant, allowing damage to be remotely inspected in areas suspected to pose a high radiation risk.

PD-100 Black Hornet

Black Hornet is termed a Nano Unmanned Air Vehicle (NUAV), in reference to its extremely small size. It went into series production in late 2012, with deliveries to the British armed forces beginning in late 2012 or early 2013.

Black Hornet fulfils the same mission as larger drones – intelligence, surveillance and reconnaissance – although its operational range is restricted by the life of its rechargeable battery, which lasts for about 25 minutes. The drone can be carried in a pocket and weighs only 16g (0.5oz);

Below: Described as a Personal Reconnaissance System, Black Hornet comes in a package with a control system and two air vehicles. Whilst it requires no skill to fly as it is simply programmed with a course to follow autonomously, using the drone to good effect requires training.

Right: Urban terrain is an extremely dangerous environment for combat personnel. The ability to scout ahead without exposing troops to sniper fire saves lives and improves mission efficiency. Using a very small drone has the advantage that the enemy might not notice it and can be caught unawares.

the complete unit, which includes two drones and a control unit, comes to less than a kilogram.

Black Hornet's short endurance is the tradeoff for creating an extremely compact drone that can operate in areas where a winged aircraft type UAV could not. The drone is designed for precise manoeuvring in a cluttered environment, such as indoors, and can thus be used to check rooms in a building for concealed enemy personnel.

A drone intended for use in this sort of urban combat environment must be capable of rapid deployment. Black Hornet can be airborne in under a minute, enabling personnel to make a quick and covert reconnaissance of an area they are about to enter, or to search for snipers without being exposed to fire. Black Hornet's rotors make virtually no noise, and it is unlikely to be spotted either optically or with any detector currently in service.

Outdoor operations are also possible. The size and shape of the drone make it resistant to the effects of wind, and it can climb vertically over walls and other obstructions – or look in and out through windows – in a manner that a winged drone cannot.

Despite being extremely small, Black Hornet carries three cameras that can produce low-light video or still images, and can be steered and zoomed in. Control is via a joystick and display unit, allowing line-of-sight direct control out to about 1000m (3280ft). Alternatively, GPS guidance can be used to create a set of waypoints through which the drone will automatically navigate.

PD-100 BLACK HORNET OPERATIONS

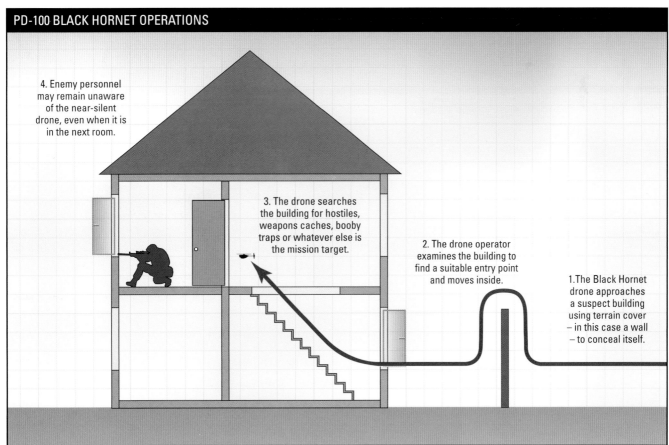

4. Enemy personnel may remain unaware of the near-silent drone, even when it is in the next room.

3. The drone searches the building for hostiles, weapons caches, booby traps or whatever else is the mission target.

2. The drone operator examines the building to find a suitable entry point and moves inside.

1. The Black Hornet drone approaches a suspect building using terrain cover – in this case a wall – to conceal itself.

Cruise missiles

Cruise missiles are, by many definitions, a form of drone with a single application. The term 'cruise missile' refers to the fact that the missile flies in the manner of an aircraft rather than being held aloft by its propulsion system or inertia as a typical missile is. A missile's fins are stabilizer and control surfaces; the wings of a cruise missile provide lift. This does create drag, which slows the missile down, but it permits the missile to remain airborne using a much lower expenditure of fuel, since it is flying rather than pushing itself up as well as forward using the thrust from its engine.

Above: Early cruise missiles were developed as a standoff delivery system for nuclear weapons, and, to this day, many people equate 'cruise missile' with 'nuclear weapon'. In fact, most cruise missiles today have conventional warheads and can be launched from aircraft, submarines and surface ships.

This means that a cruise missile can travel much greater distances than a conventional missile, and can undertake significant manoeuvres on its way to the target area. Indeed, cruise missiles can be programmed to navigate by way of a fairly complex flight path that routes them around air defence sites. They are relatively small compared to aircraft and thus harder to detect, especially when flying low, and are capable of delivering a large warhead with great precision, even over long distances.

Cruise missiles can be launched from a variety of platforms ranging from fixed or mobile ground launchers, submarines or aircraft. This mobility allows the missile to be carried close to enemy territory or to a launch point that will allow the weapon to begin its penetration of enemy airspace from an unexpected direction.

Right: Large aircraft can carry numerous missiles and launch them at widely dispersed targets. A rotating dispenser, similar in function to a revolver's cylinder, was developed to allow B-52s and other heavy bombers to carry a large load of missiles in the bomb bay.

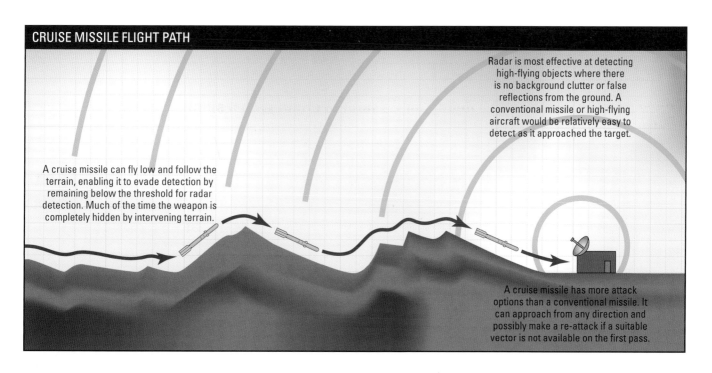

CRUISE MISSILE FLIGHT PATH

Radar is most effective at detecting high-flying objects where there is no background clutter or false reflections from the ground. A conventional missile or high-flying aircraft would be relatively easy to detect as it approached the target.

A cruise missile can fly low and follow the terrain, enabling it to evade detection by remaining below the threshold for radar detection. Much of the time the weapon is completely hidden by intervening terrain.

A cruise missile has more attack options than a conventional missile. It can approach from any direction and possibly make a re-attack if a suitable vector is not available on the first pass.

Left: The venerable B-52 Stratofortress remains an effective attack platform, largely due to its ability to deliver cruise missile strikes. The weapons carried on the underwing pylons may be supplemented by others in the bomb bay, enabling one B-52 to strike several targets without entering enemy airspace.

AGM-86

The AGM-86 ALCM (or 'Air Launched Cruise Missile') was developed as a nuclear attack platform. Originally, strategic nuclear attack was the province of bombers that had to penetrate enemy airspace to deliver their weapons. The advent of ballistic missiles that could be launched from ground bases or submarines augmented this capability, but a significant proportion of the superpowers' nuclear arsenal continued to depend upon delivery by bomber.

Improved air defences in the form of radar detection and surface-to-air missiles made the bombers' mission ever more hazardous. At one time, it was possible to fly above the engagement envelope of surface-to-air (SAM) sites, but this gap was closed by more advanced missiles. Low-level 'dash' penetration and stealthy bombers offered new possibilities, but the life expectancy of the bomber crews remained reduced.

High-flying specialist interceptor aircraft were also developed, sometimes armed with very large long-range air-to-air missiles. These specialist aircraft were poor air-superiority fighters, but this was not their intended role – the bombers had to be stopped by any means possible. Even nuclear anti-aircraft missiles were developed that would enable the interception of groups of bombers at great distances.

In this environment, it became unrealistic to imagine that a nuclear-armed bomber would be able to penetrate enemy airspace and deliver its weapons. Yet large numbers of aircraft

Above: An AGM-86 CALCM (Conventional Air-Launched Cruise Missile) is launched from the bomb bay of a B-52. Standoff strikes with cruise missiles were used to great effect at the outset of the 1991 Gulf War, striking key command, and, control centres with little chance for the enemy to respond.

existed, along with crews and support infrastructure set up to support the nuclear delivery role. If these bombers could be converted into launch platforms for nuclear missiles, they could continue in their existing role with vastly enhanced survivability.

The AGM-86 was designed to be carried by the US fleet of B-52 bombers. Up to six could be carried on underwing pylons, with eight more in the bomb bay on a rotary launcher. This gave the bombers a potent standoff attack capability and additional tactical or strategic options. Bombers could carry the missiles to a loiter point and wait for an order to attack, while retaining the capability to abort the strike right up to the last moment.

The initial missile, designated AGM-86A, was a development version. A

slightly enlarged model, AGM-86B, was put into production as a nuclear delivery system. Once launched, it was designed to fly at low level using a combination of inertial guidance and Terrain Contour Matching (TERCOM). This system compared data from the missile's radar altimeter to a pre-programmed map of the ground it would pass over, enabling closer matching of ground contours, and a generally lower flight altitude that contributed to both accuracy and the difficulty of interception.

At the time of the missile's introduction in the mid-1970s, it seemed likely that nuclear weapons would be used in any major conflict, and delivery of warheads remained the primary mission for strategic bombers. However, over time, it became apparent that conventional (i.e. non-nuclear) conflict would continue

to occur, and the decision was made to give the air-launched cruise missile a conventional strike capability.

The result was the AGM-86C and D models. Both used conventional warheads, with the AGM-86C carrying a standard explosive charge to create blast and secondary fragmentation effects. Such weapons are most effective against area targets that are not 'hardened', i.e. they are very effective against personnel, communications equipment, light structures and softskinned vehicles in a fairly large radius. Even a tank or other armoured vehicle would be destroyed if close to the point of impact, but weapons of this sort are not very effective against bunkers or underground structures.

The AGM-86D model was designed to carry a penetrator warhead. Sometimes referred to as 'bunker-busters', penetrator warheads are armoured, enabling them to punch deep into earth, rock or even concrete before detonating. To be useful, such weapons have to be able to hit a precise target. This degree

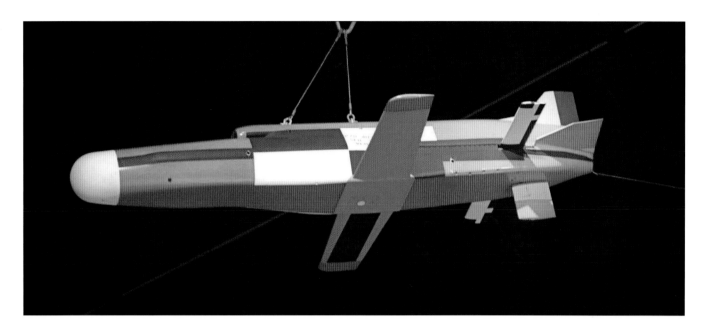

Above: The AGM-136 Tacit Rainbow missile was essentially a drone aircraft that flew itself to a position somewhere near a suspected target and then loitered, waiting for enemy air defence radar to reveal its position. The weapon was capable of making an attack/no attack decision based on the emissions it detected.

of precision was made possible by the adoption of GPS guidance.

Redesignated from ALCM to CALCM (Conventional Air-Launched Cruise Missile) the AGM-86C/D served in conflicts in Iraq from 1991 onwards and in the Balkans in 1999. Its first operational use was in 1991 at the beginning of Operation Desert Storm. B-52 bombers flew from the USA to launch points off Iraq, then delivered precise cruise missile strikes against key targets.

This extremely long flight displaced the 'Black Buck' bombing raid of the 1982 Falklands War as the longest air mission in history, in terms both of duration and distance flown. The 1982 involved Vulcan bombers from bases in the United Kingdom flying to the Falkland Islands and home again to deliver relatively small strikes using bombs and anti-radar missiles. By comparison, the 1991 raids delivered many times more ordnance with much greater effect, and were far more significant to the course of the conflict. This was mainly down to the capabilities of the final delivery system – AGM-86C CALCMs.

One variation on the concept of an air-launched cruise missile was

the AGM-136 Tacit Rainbow project, which was ultimately cancelled but demonstrated a fascinating concept. One key mission for air assets is SEAD (Suppression of Enemy Air Defences), which can be carried out by HARM (Homing Anti-Radiation Missiles). The 'radiation' referred to in this case is radar signals emitted by enemy tracking and air defence radars. When these are detected a HARM missile can be fired to home in on the signal and destroy the radar.

However, this requires that an aircraft enter the enemy's air defence zone and expose itself to attack by missiles fired from the ground. Tacit Rainbow offered an alternative: that of a PARM (Persistent Anti-Radiation Missile). Launched from an aircraft, the weapon would proceed to the general area of the suspected target and loiter there waiting for enemy radar to begin emitting. Once it did so, the AGM-136 would attack it with very little warning. The ability to carry out this mission was, of course, dependent upon

the ability of the weapon to proceed to the target area and fly itself in a holding pattern while waiting for the right emissions, then identify them and initiate an attack, before guiding itself to the target. This all required a combination of good sensors and decision-making electronics, and may have been beyond the technology of the day. The project was terminated in 1991, but remains a possibility for today's advanced and stealthy drones, clearing enemy air defences out of the way of a strike by conventional aircraft.

SPECIFICATIONS: AGM-86

Length: 6.29m (20ft 9in)
Wingspan: 3.64m (12ft)
Diameter: 62.23cm (24.5in)
Powerplant: Williams Research Corp. F-107-WR-10 turbofan engine
Weight: 1417kg (3150lb)
Maximum speed: 885mk/h (550mph)
Range: AGM-86B: 2400km (1500+ miles)

Left: Small but surprisingly sophisticated drones have become a popular recreational item in the past few years. Where radio-controlled aircraft and helicopters were something of a niche hobby, drones seem to have found a wider appeal. This may be due to the ease of flying them.

Non-Military DRONES

Non-military drones

Outside the military there are no real applications for armed drones, but other capabilities are useful in a wide range of industries. The ability of a UAV to carry a camera or thermal sensor, one of its most basic and cheapest functions, has a vast range of applications. However, there is some controversy about camera-equipped drones, which can be used to enter secluded areas and thus obtain pictures that the target would prefer were not published.

Above: Farmers have found many uses for a simple but rugged UAV. Its camera can be used to monitor conditions over a wide area much more quickly than a visual inspection would allow. A working drone also needs a solid case to protect it from the hazards of life on the farm.

From a scientific point of view, one of the main barriers to research is funding. Any project that requires remote or aerial photography will cost a great deal of money, and it can be hard to obtain funding for a project that is not highly likely to produce a good financial return over a predictable timeframe.

The use of drones can greatly reduce the costs of – and therefore make possible – a wide range of projects ranging from wildlife observation and climate monitoring to archaeology. Quite often, sites of historical interest are obvious from the air due to vegetation patterns or the arrangement of what at ground level might seem like natural features. A series of drone overflights can map a fairly large area and give archaeologists an indication of where to dig – or help them get funding to begin a more detailed exploration of the area.

Similarly, regular aerial photography can be used to observe sand dune migration and coastal erosion over a fairly wide area. This would be expensive using conventional means, but a drone-based project could be run on a relative shoestring and still collect important data.

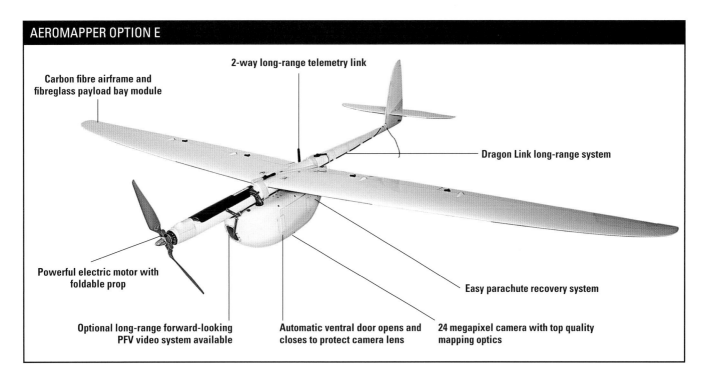

AEROMAPPER OPTION E

Carbon fibre airframe and fibreglass payload bay module

2-way long-range telemetry link

Dragon Link long-range system

Powerful electric motor with foldable prop

Easy parachute recovery system

Optional long-range forward-looking PFV video system available

Automatic ventral door opens and closes to protect camera lens

24 megapixel camera with top quality mapping optics

The same applies to studies of rivers and watercourses; repeated overflights can give an indication of how conditions change and how quickly, whereas a single set of photographs is at best a snapshot of conditions on a given day.

Law Enforcement

Cameras are also highly useful for security and law enforcement purposes. It may be far more cost-effective to monitor a large estate or secured area using a drone overflight and a few fixed cameras than to employ a large security force, and, where a remote but not commonly used area is being monitored, it may be sufficient to send a drone over once in a while rather than visit in person.

This approach might be useful to landowners concerned about squatters or other unwelcome occupants moving onto their land, or misuse for other reasons. Tyre tracks, flattened areas of vegetation and damaged walls and fences can provide clues that someone is using the land without permission. It may be that legal attempts to

remove unwelcome occupants will be strengthened by a quick response, whereas those who have been living on a quiet corner of a remote area for some time may be more difficult to shift.

Drone cameras can be used in law enforcement, from recording traffic offences to monitoring an unruly crowd to an all-out riot. Airborne cameras can give the on-scene controllers a 'big picture' of what is happening, as well as zooming in on points of interest or recording incidents for later use as evidence. An airborne vantage point can be used to direct security personnel or police officers to an incident or to track suspects after they leave the immediate area.

Increasingly, video or photographic evidence is necessary in a court case. This is an inevitable consequence of increasing availability; as it becomes more common, cases without it seem weaker. Thus the ability to monitor an incident ensures that evidence is available if needed. The fact that a police drone is nearby may also act as a deterrent, and protects against

allegations made against police or security personnel. Drones are also useful in other filming applications, such as television and movies. An airborne camera can shoot from angles that might not be available to a ground-based crew without the need to hire a helicopter or construct a scaffold. This might not be much of a consideration to big-budget movie companies, but for smaller productions the capability might create options that would otherwise not exist.

Commercial Transportation

Drones are also useful in a commercial transportation capacity. Already small packages can be delivered in this way; in time, larger consignments will become practicable. For deliveries to remote areas, the primary problems are technological, but there are also legal barriers to the operation of delivery drones in an urban environment.

It might be convenient to have large packages or even the entire stock for a store delivered to a backyard or rooftop, and this does away with the problems of driving a large delivery vehicle through traffic and parking it near the destination. However, the urban environment is cluttered with hazards – even before the possibility of encountering another drone is considered. The technological problems in terms of collision avoidance and safe operation are significant, and even if they can be overcome there will be legal questions to answer.

UAVs IN ACTION

Drones can be used on a wide variety of missions, many of them involving lower costs and lower risk than traditional methods. Here are just some ways that drones can be used:

- Archaeological surveying
- Security – surveillance, crowd monitoring and control
- Law enforcement – surveillance, traffic monitoring, search and rescue, tracking a fleeing criminal
- Mapping for forestry studies
- Aerial inspection and monitoring
- Data capturing in areas with limited access or affected by environmental disasters
- Meteorology – storm monitoring, mapping glaciers, general data collection
- Humanitarian missions – medication and vaccine delivery in areas difficult to access
- Agriculture and livestock data gathering – pest detection in crops, agricultural yields, pesticide spraying, livestock counting
- Environmental surveillance – illegal mining, illegal woodcutting, invasion of protected areas, over-hunting
- Firefighting – forest and fire monitoring, risk management
- Wildlife conservation – tracking animals

Above: Law enforcement agencies have been using helicopters for many years, but they absorb a large chunk of the agency's budget and cannot provide constant coverage. Smaller UAV versions, such as Scorpio 30, offer many of the capabilities of a manned helicopter, at a fraction of the operating cost.

It seems likely that licensing will be a barrier to most potential urban delivery drone operators, along with the expense of buying and maintaining a suitably reliable drone. This option is only likely to be viable for those with a lot of money to spend, at least for the foreseeable future.

Disaster Zones

Automated vehicles may see significant use in disaster management as observation platforms, and possibly for supply delivery and casualty-evacuation missions. Thermal and visual cameras are invaluable in monitoring a disaster and looking for survivors – again, an airborne vantage point can be extremely useful in coordinating a response. It is possible that fire trucks and some police response vehicles will carry drones in the future, enabling the users to quickly assess a situation before sending personnel into potential danger. Small reconnaissance drones are very useful to small military units, and they may be equally effective in assisting fire crews and other responders.

While routine deliveries of furniture and similar large items might be some way in the future, in a disaster situation an automated delivery system may be viable. Delivery drones might be programmed to go to the location of the on-site command vehicle, delivering bulky supplies that cannot be carried aboard a fire truck or ambulance. Drones might also be used to drop water on forest fires. Since they can fly endless missions without pilot fatigue becoming an issue, transport drones may find a use in such situations, maintaining a constant high tempo of operations around the clock.

Underwater Drones

Recreational drone flying is increasingly popular, and there are underwater craft that fulfil a similar purpose. Some are quite expensive and have 'serious' uses as well, but it is increasingly possible to explore a river from underwater, then upload the images to the internet.

Most underwater drones are too costly for recreational use, however. Military and law-enforcement uses include harbour protection and minehunting, while scientific and commercial operators can use underwater drones for seabed mapping, wildlife monitoring, pipeline inspections and all manner of other tasks that would previously have required a team of divers.

The non-military applications of drones are thus potentially even wider than in the military sector, but without military budgets to drive research, progress is not as quick. However, technology developed for military applications often finds its way into other applications. Last year's combat drone may be next year's automated research plane.

Below: The Parrot AR drone is intended for recreational use. It is controlled from a hand-held console or phone. Images from the onboard camera are streamed back to the user's phone, allowing video or still images to be obtained and shared.

Below: The first regular UAV delivery service was established in 2014. A small quadcopter drone carries packages to the island of Juist in the North Sea, where a locally based courier receives them and makes the final delivery to the island's 2000 inhabitants.

NASA drones

NASA has many uses for unmanned vehicles. Often an experimental aircraft will be tested under remote control or in automated flight before risking a human crew, and there are some concepts that are sufficiently risky that a drone is the only safe option. However, some drones are used for entirely different reasons. An unmanned aircraft can stay aloft for a period far beyond the endurance of a human crew, permitting observations to be made over an extended duration.

Above: NASA Pathfinder was developed as part of a project to create an alternative to expensive high-altitude aircraft or satellites. It carried instruments to altitudes that could previously only be obtained using rocket propulsion and showed that a solar powered aircraft could remain there for a long period.

An unmanned aircraft can often be used as a testbed for electronics or sensor systems more cheaply than a manned aircraft. A drone need only be big enough to carry the instrument to be tested, and need only use enough fuel to lift a small weight. Airborne testing can therefore be done far more cheaply than it would by installing the system in a full-sized aircraft.

There are also some drones that can go places that manned aircraft cannot, at least not without enormous cost. Extremely high altitudes can normally only be maintained by specialist aircraft, which are hugely expensive. An unmanned vehicle, with no need to carry people and the systems they need to support them at high altitude, can be built much more lightly. That, in turn, translates to less weight to lift, allowing the craft to function with smaller engines and/or less fuel. This again feeds back into decreasing weight.

Some of the drones used by NASA take this concept to an extreme, creating super-lightweight craft that can operate for long periods at altitudes previously only attainable by the brute-force method of using rocket propulsion to lift an expensive and technologically sophisticated vehicle whose mission payload might not be significantly greater.

Pathfinder/Pathfinder Plus

Pathfinder had its origins in a military project initiated by what was then the US Ballistic Missile Defence Organization

(now Missile Defence Agency, or MDA). It has long been accepted that the best time to shoot down ballistic missiles is when they are launched rather than in mid-flight or as they approach the target. Not only is targeting simpler, but the wreckage of the weapon and its warhead will fall on the territory of the launching

Below: Pathfinder was modified by adding a longer central wing section, creating Pathfinder Plus. With a longer wing and improved solar cells, Pathfinder Plus was brought back up to its original eight-motor configuration. These were more powerful than the original engines, which in turn increased payload capacity.

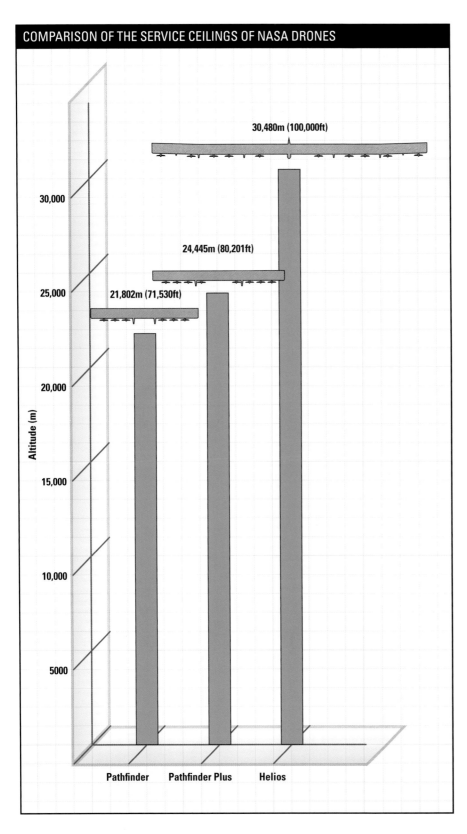

COMPARISON OF THE SERVICE CEILINGS OF NASA DRONES

30,480m (100,000ft)

24,445m (80,201ft)

21,802m (71,530ft)

Altitude (m)

30,000

25,000

20,000

15,000

10,000

5000

Pathfinder   Pathfinder Plus   Helios

nation rather than the target. To this end, a high-altitude long-endurance drone was developed, named Responsive Aircraft Program for Theater Operations (RAPTOR). RAPTOR drones were to loiter over enemy territory using an extensive package of thermal detection equipment to scan for the hot exhaust of a missile plume. The missile would then be engaged by a hypervelocity Talon missile carried aboard RAPTOR and destroyed within moments of launch.

To support the Talon missile-equipped RAPTOR drones, a support UAV named RAPTOR/Pathfinder was developed from an earlier project. RAPTOR/Pathfinder was powered by solar cells on the upper wing surfaces, but could not generate enough power to remain at altitude all night. The project was not a success, and the two drones went to NASA for experimental purposes. RAPTOR/Talon was mainly used as an engine testbed, as its conventional configuration made it unsuited to high-altitude scientific operations, but RAPTOR/Pathfinder found a new lease of life.

Now named Pathfinder, the former missile-defence drone was stripped of two of its eight engines, but otherwise unchanged. In 1997, it set a new world record for high-altitude flight by a propeller-driven aircraft and was used by NASA for a variety of experiments before being upgraded to create Pathfinder Plus. Pathfinder Plus smashed Pathfinder's 21,802m (71,530ft) altitude record, climbing to 24,445m (80,201ft).

SPECIFICATIONS: PATHFINDER

Length: 3.6m (12ft)
Wingspan: 29.5m (98ft 4in)
Gross weight: Approx. 252kg (560lb)
Airspeed: Approx. 27–32km/h (17–20mph)
    cruise
Ceiling: 21,802m (71,530ft)
Powerplant: 6 electric motors
Endurance: About 14–15 hours, daylight
    limited with 2–5 hours on backup batteries

COMPARISON OF WINGSPAN OF NASA DRONES

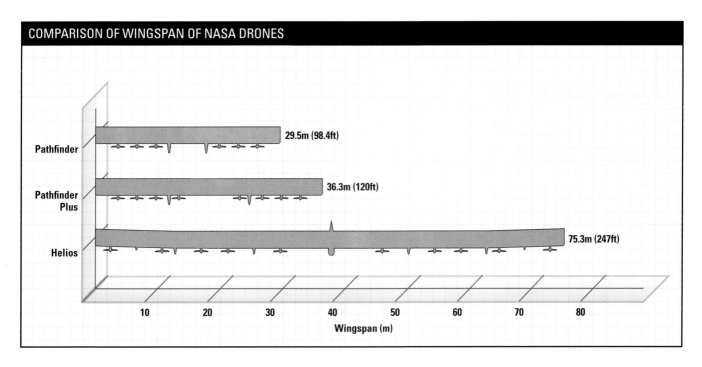

Pathfinder — 29.5m (98.4ft)

Pathfinder Plus — 36.3m (120ft)

Helios — 75.3m (247ft)

10   20   30   40   50   60   70   80

Wingspan (m)

Pathfinder was a lightweight and somewhat fragile aircraft propelled by electric motors. The original design had eight motors. Two were deleted when NASA took over Pathfinder, but were restored during the upgrade to Pathfinder Plus; they were ultimately removed again to make room for additional sensors. The motors were initially battery-powered, but from 1996 solar panels were added on the top surface of the wing. The wing itself makes up most of the craft, with two short vertical sections projecting form the underside of the wing, plus a pair of gondolas to carry equipment.

SPECIFICATIONS: PATHFINDER PLUS

Length: 3.6m (12ft)
Wingspan: 36.3m (121ft)
Gross weight: Approx. 315kg (700lb)
Airspeed: Approx. 27–32km/h (17–20 mph) cruise
Ceiling: 24,445m (80,201ft)
Powerplant: Eight electric motors
Endurance: About 14 to 15 hours, daylight limited with two to five hours on backup batteries

Pathfinder was very much an experimental project to see what could be done with solar-powered drones and to investigate the technologies required. Uses for such drones include high-altitude atmospheric research and the capability to act as an airborne communications relay point, acting much like a communications satellite, but at a fraction of the cost. Lessons learned with Pathfinder were implemented on the next incarnation of the concept, named Centurion.

Centurion/Helios

Pathfinder Plus was followed by a larger version of the same general design, named Centurion. It was part of a NASA project named ERAST (Environmental Research Aircraft and Sensor Technology) that was designed to prove the concept of and develop the technology for atmospheric pseudo-satellites that could be used for communications purposes. It resembled a larger version of Pathfinder, but incorporated some redesign to increase carrying capacity. With a larger wing, 14 engines to Pathfinder's eight

and four equipment gondolas, Centurion was designed to reach a milestone in the ERAST project by climbing to a height of 30,480m (100,000ft).

In the event, launch was delayed on the test date and the Centurion UAV did not have enough hours of sunlight to complete its climb; once darkness fell, solar cells were, of course, useless and, at that point it had no means to store energy. It did reach 29,413m (96,500ft), which was something of a mixed blessing. On the one hand it was an incredible achievement and set new records. On the other, it was short of the project's goal, but close enough to it that a second attempt was considered not to be cost-effective.

Development work on Centurion continued until the aircraft was destroyed in 2003. Atmospheric conditions caused the fragile craft to become distorted and to experience an oscillation of its long wing. Excessive airflow over the wing caused sections of solar panel and parts of the upper wing surface to become detached. Centurion broke up as it fell, although most

COMPARISON OF THE GROSS WEIGHTS OF NASA DRONES

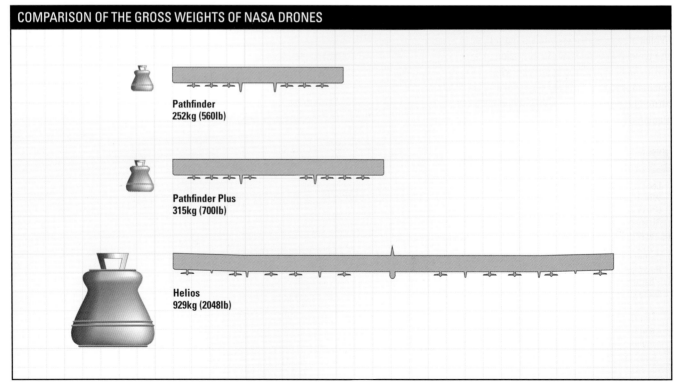

Pathfinder
252kg (560lb)

Pathfinder Plus
315kg (700lb)

Helios
929kg (2048lb)

Left: NASA Centurion was an evolutionary step forward in the Pathfinder/Pathfinder Plus family, using the same general design, but with a longer wing made up of more segments. These supported 14 motors and four underwing payload pods instead of the eight and two of Pathfinder Plus.

components were recovered. Pathfinder Plus remained in service for another two years after the loss of Centurion, and finally retired in 2005. However, Centurion was always intended to be no more than the prototype for a more ambitious project named Helios, which went ahead with new and larger versions. Helios uses the same general design concept as its predecessors, consisting mainly of a long wing with pods under the joins between each of the wing sections.

Helios has no rudder. It turns horizontally by slightly increasing the power output of the motors on one wing. This method also allows control over pitch, since in flight the natural flexing

Right: The delicate balance between strength and lightness is very apparent here. Whilst all aircraft wings bend in flight due to the weight hanging from them, Helios took this to an extreme. In June 2003, turbulence encountered on a test flight caused a structural failure and Helios crashed in the Pacific Ocean.

or bowing of the wing means that the outer engines are higher than those in a more inboard position. Accelerating the higher motors pitches the nose down a little – accelerating the lower ones has the opposite effect. Helios also has elevators on the trailing edge of the wing to provide additional pitch control.

The target altitude of 30,480m (100,000ft) has still not been exceeded, meaning that Centurion continues to hold the world record for altitude

achieved by any non-rocket powered vehicle. The project has shown that it is possible to create an extremely high-altitude UAV suggesting that the 'atmospheric satellite' concept is workable – and there is another important implication of successful flight at this altitude. At 30,480m (100,000ft), air density is similar to the atmospheric density of Mars. Helios may have proven that it will someday be possible to explore Mars using a similar vehicle.

SPECIFICATIONS: HELIOS

Length: 3.66m (12ft)
Wingspan: 75.3m (247ft)
Gross weight: Up to 929kg (2048lb)
Airspeed: From 31–43km/h (19 – 27mph) cruise at low altitudes, up to 274km/h (170mph) ground speed at extreme altitude
Ceiling: 30,480 (100,000ft)
Powerplant: Bi-facial solar cells. Lithium battery pack backup power
Endurance: Daylight hours, plus up to 5 hours of flight after dark on storage batteries. When equipped with a supplemental electrical energy system for night-time flight, from several days to several months

Agriculture and wildlife drones

Scientists studying wildlife or natural phenomena, such as monitoring the spread of a river delta, find aerial pictures extremely useful. Indeed, they can be the only way to obtain images of some regions without extremely lengthy and sometimes hazardous expeditions on the ground. Wildlife may, of course, shy away from a group of people, so a ground expedition might not be a particularly good way to observe in any case.

Above: Aerial photographs from a UAV, such as Precision Hawk, can quickly tell a farmer where there are dry areas or perhaps diseased crops. Many warning signs would not be visible at ground level until it was too late to prevent considerable damage.

The use of a UAV solves a number of problems for the ecological scientist. Since most drones are very quiet and likely to be mistaken for birds, it is more likely that they will capture images of shy wildlife than a ground-based camera team. This will give a more realistic picture of numbers, habitats and behaviours as they really are when humans are not around.

A drone can also be used to reach and observe remote areas that otherwise would require too much effort, and to repeatedly photograph the same region. This would be prohibitively expensive with conventional aircraft, but can be carried out on a near-constant basis with drones. Thus rather than a single snapshot of the area on the day an overflight was made, researchers can compare photos taken monthly, weekly or even daily.

This is very useful when observing the effects of climate conditions throughout the year, and not only in remote places of interest primarily to well-funded scientists. Farmers and those operating in other industries that impact the local environment and are affected by it can also monitor conditions using a UAV.

A UAV can inspect a pipeline for leaks, or investigate what effect it is having on the surroundings, more efficiently than sending a crew out to physically observe – even if that is possible. A farmer can likewise obtain the 'big picture' of what is happening on his land much more efficiently with an airborne camera than by driving around, even if he could see enough from ground level.

UAV images are highly useful in spotting water distribution, as drier vegetation changes colour. Remedial action can thus be taken before part of a crop is lost. Direct action can also be taken with drones, such as using automated helicopters to spray crops rather than hiring a crop duster aircraft or attempting to do it by hand. It is possible that, in future, roadside vegetation may be kept down by automated spraying with herbicide, reducing exposure of the operator to potentially harmful chemicals. However, this would require strict controls before an unmanned vehicle could be sent out carrying a tank of poison and the means to spray it wherever its potential malfunction dictated.

CropCam

CropCam, as its name suggests, is designed to meet the needs of farmers and those needing to study environmental conditions from the air. It takes the form of a small hand-launched glider-like aircraft with a digital camera on board.

Below: CropCam can be manually controlled like a traditional radio-controlled aircraft, or can fly a preprogramed course set up on the operator's laptop. The UAV is reportedly able to survive encounters with trees, and has continued to function after very rough landings in difficult conditions – including coming down in a river.

Once launched, CropCam has an endurance of about an hour, which is sufficient to photograph a fairly large area. Varying the height at which the UAV flies will effectively alter the resolution of images, and allows either a wide-area picture to be built up or a fairly precise set of photographs to be taken. The area to be photographed can be set before flight, with onboard GPS guidance used to fly a pattern over the target area and then return to the takeoff point. An overflight pattern can be repeated numerous times to build up a picture of the situation over a period or at a certain time each year.

Satellite images are sometimes available, as are those taken from conventional aircraft, but there is no guarantee that these pictures will be up to date, even if they are available at all – unless, of course, the landowner wishes to spend quite a lot of money on a photographic overflight. The advantage of using a drone is that images can be obtained on almost any day, providing the weather is reasonably good, and at a relatively low cost.

In addition to general monitoring of how crops and other plants are doing, CropCam can be used to watch for problems before they happen, such as disease, water shortage or over-grazing. Drones of this sort also have applications for the logging industry and potentially also fisheries and waterways. Clogged watercourses or excessive vegetation in or around the water can be detected as easily as problems with foliage and crops on land, without having to make a boat trip or struggle through undeveloped and perhaps overgrown land to the waterside.

The term 'precision agriculture' has been coined in recent years, and the increasing availability of inexpensive drones has been a major factor in obtaining maximum yields and dealing with problems early. The use of UAVs of this sort is likely to expand over time since the benefits greatly outweigh the modest costs.

Above: CropCam and other drones can fly over obstacles, but they have to descend to land. When flying a UAV with automatic return to takeoff point capability, it is worth considering what direction it will approach from and any obstacles there may be. This is especially important when operating in a forestry or watercourse-monitoring environment.

MAJA

The MAJA UAV is a small aerial photography platform suitable for environmental monitoring, wildlife observation and similar applications. It uses an efficient design with a blocky fuselage that maximizes payload space. It can carry a load equal to its own weight of 1.5kg (3.3lb). This is typically two cameras plus batteries and guidance system. Access to these components is via the top of the fuselage, which opens up along much of its length.

MAJA has a straight high wing and a pusher propeller, all of which are of robust construction. The drone is designed to be used in remote areas where it may encounter all manner of hazards. It can land automatically, but it is not able to detect obstructions, so a certain ruggedness is necessary in case the drone encounters something hard as it comes down.

Drones have seen increasing use in conservation and environmental research

projects in recent years, and, while MAJA is somewhat more expensive than some of its rivals, this tradeoff has proved worthwhile in terms of keeping the cameras flying in sometimes difficult conditions. Obtaining a replacement drone might prove a challenge when observing orangutans in the forests of Sumatra, which is one of the roles this drone has filled.

Given that the drone marketplace is expanding, it is hardly surprising that after-market upgrade kits are becoming available for modular drones like MAJA. Wing kits, upgraded motors and

SPECIFICATIONS: CROPCAM

Length: 1.2m (4ft)
Wingspan: 2.4m (8ft)
Weight: 2.7kg (6lb)
Powerplant: Electric Axi Brushless
Ceiling: Can be adjusted to suit regulations in each country 122–671m (400–2200ft)
Duration: 55 minutes
Average speed: 60km/h (37.2mph)

more powerful batteries are beginning to appear, not just from the drone manufacturers, but also from third-party companies who make spares and upgrade kits for a variety of other manufacturers' products.

It may be that in time the drone marketplace will begin to resemble many other marketplaces, such as the automotive after-market industry or the computers sector, with a vast array of generic sensors, control devices and add-ons marketed for those who want a bit more performance or simply to engage in the one-upmanship that has driven certain other industries.

Above: The design of small UAVs, like MAJA, is intended to maximize their utility by creating as much usable internal space as possible and making it simple to swap payloads. A modular design also makes replacing damaged components much easier, which can be important for a working drone.

Underwater drones

The underwater environment can be extremely hazardous. Even in shallow water there are always dangers, such as submerged rocks, tides and currents. Passing boats can endanger a diver in shallow water, and wildlife can also be a threat. While the danger of being attacked by a shark or a squid is rather less than most of us imagine, there is always the possibility that wildlife can contribute to a diver's troubles. Inquisitive creatures can be a real nuisance, and even a passing shoal of fish can obstruct vision and hide a hazard.

Above: Deep Drone 8000 gets its name from the fact that it can operate in 8000ft (2848m) of water. At this depth, a submarine or ROV is subjected to almost 250 times atmospheric pressure. Protecting humans against this level of pressure is far more difficult than creating a drone that can withstand it.

It is the environment that is the primary hazard. Water pressure increases rapidly with depth, which places severe limitations on the depth at which a diver can safely operate without specialist equipment and training. Even then, there are significant risks to be dealt with during what are rightly referred to as deep dives, but which are extremely shallow compared to the depth of oceanic waters.

Specially constructed submarines are required to convey humans to the bottom of ocean trenches or even the ocean-bottom plains. These are enormously expensive and require massive support – and they are necessarily limited in both their capabilities and their endurance.

Underwater vehicles, whether autonomous (Autonomous Underwater Vehicle, AUV) or remotely operated (Remotely Operated Vehicle, ROV), do not need to support a human crew. This means that they do not need to withstand a huge pressure differential between the environment outside and that inside. They can thus be built much more lightly and are less susceptible to catastrophic

Right: Small underwater ROVs, like HydroView, have a number of recreational applications, from fish-watching to exploring an area that the user is not likely to ever be able to visit. They also have many 'serious' uses, such as scouting an area before divers enter it or conducting underwater inspections.

damage. They can also, of course, be smaller than submarines because they do not need to have space for the crew to work, nor for the support equipment required to keep them alive.

Hydroview

HydroView is a remotely-operated vehicle rather than a true drone in the sense of being autonomous. It can be operated from a laptop or tablet, requiring no training, and feeds video back to the user's display. It might be considered – or used as – nothing more than an expensive toy by some, but small ROVs of this sort can be used to carry out a range of useful functions.

HydroView consists of a central equipment and electronics section, flanked by the propulsion units. Its camera can stream video images or take stills, and, in addition to recreational uses, can undertake a number of safety-related or money-saving tasks.

Underwater inspections normally require a diver to enter the water and carry out the task manually, which requires trained personnel and expensive equipment. A boat can be dry-docked or an area drained of water in some cases, but this is a lengthy undertaking and not always practicable. However, an ROV can be used to carry out the inspection with less preparation time and no risk.

HydroView can enter small spaces that could not accommodate a diver, or cluttered areas that would not be safe to enter. It can be used to search for objects underwater, inspect the underside of a boat or check a sea wall for cracks. The ability to deploy an ROV

at very short notice can be useful when a situation occurs that might or might not be hazardous. An ROV could quickly be used to see what was fouling a boat's propeller or to investigate an object just under the water that might not be safe for a vessel to approach.

Thus small, simple ROVs have a number of applications that can save time and money, or increase safety, as well as obtaining some interesting pictures to share on social media sites.

At present this requires direct control, but it may not be long before automated underwater drones are in use to carry out routine and frequent inspections of underwater cables, flood defences and other areas that would normally require a diver to access. This translates to more and cheaper inspections and the chance to spot problems before they become serous, which could, in the long term, prevent a more serious situation from ever occurring.

Deep Trekker

Deep Trekker consists of a central section containing a stabilized camera, plus propulsion units on the sides. It is powered by rechargeable batteries and operated using a simple hand-held controller. The ROV operates at the end of a tether, which helps prevent losses due to unexpected events or a discharged battery, and can be used to lower Deep Trekker into the water from high above, such as down the side of a ship or off a bridge.

Deep Trekker was developed for a number of applications, including acting as a preliminary reconnaissance unit for divers. While the dive team is preparing its equipment, the ROV can be used to check the water entry point is safe or to reconnoitre the target. In many cases, it can conduct an inspection without needing to bring in divers at all.

This capability is of interest to commercial users, who can inspect pipes, water tanks and underwater infrastructure, and security operators who may use the ROV to check under vessels as they enter or leave a harbour. Deep Trekker is also used for environmental monitoring, and in aquaculture where it can check the condition of both nets and fish.

Deep Trekker's onboard camera can track within the main hull, enabling the user to change viewing angle without having to undertake complex manoeuvres. The camera's aim point is matched by a small floodlight, enabling the ROV to operate in fairly dark water. It can also carry a range of accessories, including a sonar set and a grabber for retrieving objects.

Left: ROVs, like Deep Trekker, can carry out operations that would normally require divers, such as inspection of underwater cables, pipes and obstructions. This not only reduces risk to personnel, but also saves time and money – a drone can be in and out of the water in less time than a dive team.

Automatic systems can be set to maintain depth or heading, which is a step towards autonomous operations. However, manual control is still required and is assisted by an array of sensors.

Some report on external conditions such as water temperature and depth; others are internal monitors indicating battery status, pitch and roll situation and camera angle.

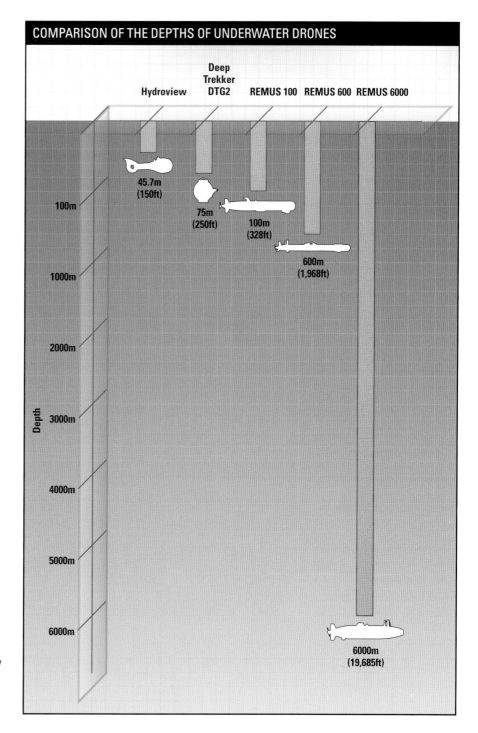

COMPARISON OF THE DEPTHS OF UNDERWATER DRONES

Hydroview
Deep Trekker DTG2
REMUS 100 REMUS 600 REMUS 6000

45.7m
(150ft)

75m
(250ft)

100m
(328ft)

600m
(1,968ft)

6000m
(19,685ft)

Depth

100m
1000m
2000m
3000m
4000m
5000m
6000m

Deep Trekker, and ROVs like it, are doing in the underwater environment what inexpensive civilian drones are doing in the air. Some are used as toys, but many fulfil useful functions that reduce the risks to personnel, or the number of man-hours required to carry out a given task. As ROVs find an increasing role in commercial, security and environmental applications it is likely that costs will go down and capabilities will continue to increase.

VideoRay

VideoRay is designed along the lines of a miniature submarine, with a central pressure hull containing the payload. It can operate at depths of up to 300m (984ft). By way of comparison, the Professional Association of Diving Instructors considers depths of 18–30m (60–100ft) as a deep dive. Going much deeper than this requires complex gas mixes, or an advanced atmospheric diving suit that is to some

Below: VideoRay 4 is available as a range of configurations and can be tailored to the application at hand. Lightweight and more rugged versions of the control system are available. Which is most suitable depends on the environment the user expects to operate in, or how mobile he needs to be.

SPECIFICATIONS: DEEP TREKKER DTG2	
Length: 27.9cm (11in) **Width:** 32.5cm (12.8in) **Height:** 25.8cm (10.2in)	**Weight in air:** 8.5kg (18.7lb) **Depth:** 75m (250ft)– 125m (410ft)

extent a human-shaped submarine. Sending divers down to work or conduct inspections at significant depths is hazardous and requires advanced training. Some tasks, of course, require this, but the use of an ROV enables divers to be deployed less often, and to benefit from an initial reconnaissance of the dive area.

VideoRay is propelled by two thrusters at the rear. The propellers can be reversed to move backwards, or one can be reversed to turn the ROV around in place. It operates at the end of a tether that can be used to retrieve the ROV in the event of a malfunction. The tether can also be used to recover objects that are too heavy for the ROV to lift on its own, by securing them with a grabber attached to the VideoRay ROV and then simply hauling on the line. A range of tethers, with different buoyancy characteristics, are available to suit various needs.

Among the most significant difficulties in underwater operations are surface conditions (e.g. waves and wind) that can make launching a larger submersible or deploying divers very hazardous and dangerous underwater currents. Exploration of wrecks can also be hazardous for divers. A small UAV can be deployed if need be by throwing it over the side of a parent vessel, and, provided it has a good thrust-to-weight and thrust-to-drag ratio, it will be able to operate in quite strong currents.

In addition to cameras, VideoRay can carry a sonar unit. This has many applications, particularly in deep and dark water. It has been successfully used by law enforcement agencies to search for bodies and evidence concealed in water,

by harbour security agencies and by the military to locate dangerous ordnance sunk in water.

VideoRay ROVs have seen extensive use in scientific applications, notable in regions such as Antarctica where conditions can be very hazardous for divers. ROVS have also been used to film and map changes in reefs, and to observe, fish for both scientific and aquaculture purposes. Indeed, ROVs of this sort can be useful in almost any industry or endeavour, either as an

assistant to divers, or to carry out routine tasks and inspections that can be more efficiently performed by a ROV.

REMUS

REMUS is a torpedo-shaped Autonomous Underwater Vehicle (AUV) designed to carry a variety of payloads. It is not a single vehicle, but a 'family' using some components in common, with different diameters and internal capacities. These range from the fairly compact REMUS 100, which is designed

Below: The REMUS series are Autonomous Underwater Vehicles (AUVs) rather than Remotely Operated Vehicles (ROVs). REMUS 100 is designed to be easily transportable and can be carried by two people. Other AUVs in the same family are capable of deeper-water operations, but are larger and heavier.

SPECIFICATIONS: VIDEORAY PRO 3

Length: 37.5cm (14.75in)
Width: 28.9cm (11.4in)
Height: 22.3cm 8.75in)
Weight: 6.1kg (13.5lb) [with full ballast set]
Depth: 305m (1000ft)

Above: REMUS 600 is designed for deep-water operations or long-duration missions in fairly shallow water. It uses the same data transfer and control system as the rest of the REMUS family and can be equipped with a wide range of underwater sensors selected by the operator.

SPECIFICATIONS: REMUS 100

Vehicle diameter: 19cm (7.5in)
Vehicle length: 160cm (63in)
Weight in air: 38.5kg (85lb)
Maximum operating depth: 100m (328ft)
Endurance: Typical mission endurance is 8–10 hours dependent on speed and sensor configuration, operating environment and mission program

for use in coastal waters up to 100m (328ft) deep, to the REMUS 6000, which is capable of very deep-water operations.

With any underwater vehicle, transport to the operating area and deployment of the AUV presents a problem for the operators. The REMUS system includes a launch and recovery system that can be set up on the stern or side of a wide range of vessels. It is designed to be brought into operation fairly quickly, which is an advantage for scientific or commercial AUV operators who rely on chartered vessels for transport.

Underwater navigation can also pose challenges. The REMUS family can use a system of acoustic transducers to create a sort of 'underwater GPS'

by measuring the distance between the AUV and each of the transducers. These can be hull mounted on the parent vessel, giving a relative position or placed in the operating area to give an absolute position reference within the working area.

REMUS can carry an array of equipment, including cameras and a variety of sonar systems adapted for different applications. Sonar can be used as an altimeter (giving height above the seabed) for seabed mapping, to indicate water conditions, or to search for specific objects, as well as for navigation. Synthetic aperture sonar can be used in much the same manner as the radar equivalent, utilizing the

movement of the AUV to sweep an area and build up a detailed picture over time.

REMUS AUVs have been used for underwater search operations, for example, looking for air crash wreckage, for harbour security, and for a variety of scientific applications, including seabed surveys, environmental monitoring and hydrographic research. The US Navy uses AUVs for minehunting, which is much safer than the more traditional method of sending a specially configured vessel (one less likely to set

SPECIFICATIONS: REMUS 6000

Vehicle diameter: 66cm (26in)
Vehicle length: 3.99m (157in)
Weight in air: 240kg (530lb)
Maximum operating depth: 6000m (19,685ft) ;
4000m (11,244ft) configuration also available)
Endurance: Typical mission duration of 16
hours

Above: The REMUS family of AUVs have proven effective in naval minehunting operations. They can be deployed from a range of craft, including an unmanned vessel that the Royal Navy is experimenting with. The combination of unmanned parent craft and AUVs further removes personnel from the risks inherent in counter-mine operations.

off mines) into a suspected mined area to conduct a search.

Although humans have been sailing the oceans for centuries, we know relatively little about the underwater environment, especially in deep water. New species and natural phenomena are being discovered all the time, and, as the search for resources becomes ever more intense, the ability to explore the ocean beds will increase in importance. This can be accomplished at lower cost and risk with AUVs than by sending a submarine down into the depths. Operations and the variety of equipment available will likely expand rapidly over the coming years.

REMUS APPLICATIONS

APPLICATIONS	REMUS 100	REMUS 600	REMUS 6000
OFFSHORE (OIL & GAS)			
Baseline Environmental Assessment	●		
Geological Survey			
Pipeline Survey	●		
Debris/Clearance Survey	●		
ENVIRONMENTAL MONITORING			
Emergency Response	●	●	●
Water Quality	●	●	●
Ecosystem Assessment	●	●	●
HYDROGRAPHY			
Route Survey	●	●	
Habitat Mapping			
Deep Sea Mining			●
Charting	●	●	
EEZ Survey		●	●
Pre/Post Dredging Survey	●	●	
SEARCH & RECOVERY			
Asset Location	●	●	●
Marine Archaeology	●	●	●

Experimental unmanned vehicles

Aeronautics is a dangerous field in which to conduct experimentation. Ground vehicles do crash and watercraft can sink, but a systems failure or control loss at high altitude and great speed may be a lot harder to survive. There is also the factor that, while much can be done with models and wind tunnels, once flight testing has begun there is really no way to avoid the need to be moving fast enough to fly and doing so at a significant distance above the ground. Thus problems with an experimental aircraft tend to be catastrophic, and test pilots are highly respected for the risks they take.

Above: The Fieseler Fi 103R, better known by the project codename Reichenberg, was a manned version of the V-1 flying bomb. In theory, the pilot had a chance to bale out once the final approach to the target was underway, but his chances of survival were rather slim.

Historically, some unmanned vehicles have flown with a pilot aboard during tests, either to iron out problems or as a safety measure during testing. The German V-1, for example, suffered significant problems during testing that were eventually investigated by a test pilot riding a V-1 in a makeshift control station. Having obtained first-hand data on how the vehicle was behaving and why it was going out of control, the pilot could suggest appropriate measures to remedy the problem.

Where possible, the opposite approach is used. Unmanned vehicles are used to investigate the flight characteristics of an experimental project, or to conduct investigation into flight-related phenomena. This is particularly hazardous around the 'sound barrier', a speed that varies depending on height and air temperature. Sound propagates through air at a certain rate due to the interaction of air molecules, meaning that the speed of sound is determined by physical laws governing the behaviour of those molecules.

When an aircraft attempts to push through the air at faster than this natural

Above: In 1947, when the Bell X-1 became the first aircraft to exceed the speed of sound, there was no option but to use a manned aircraft. Today, unmanned research planes offer an alternative, especially for risky initial flights of new-concept aircraft whose transonic characteristics might not have been correctly predicted.

Right: The X-51 is a technology demonstrator rather than a true aircraft, and is carried aloft under the wing of a B-52 bomber. X-51 is intended to investigate hypersonic flight characteristics, after which the knowledge gained will be incorporated into weapons like HSSW (High Speed Strike Weapon) and possibly other hypersonic aircraft projects.

maximum speed, phenomena related to airflow become much more extreme. Air is always displaced by an aircraft's wings and fuselage, but, as the sound barrier is approached, the airflow over the aircraft and its control surfaces changes. This can cause immense structural stresses,

Above: X-51, sometimes known as WaveRider, was developed to investigate hypersonic flight. At such high speeds, newly developed materials are needed, as compression heating can destroy conventional aircraft components. The behaviour of an aircraft in these conditions is significantly different to its flight characteristics at subsonic and transonic speeds.

vibration, even a reversal of controls. As speed increases beyond the speed of sound, heating effects also become a significant factor.

An aircraft moving at supersonic speeds that begins to oscillate will suddenly present a less aerodynamic aspect to the air. This can tear off critical components, such as control surfaces or even wings, and, in extreme cases, can have an effect not dissimilar to running into a cliff face. It is obviously best that if this occurs, it does not happen when there is a crew aboard. For this reason,

unmanned air vehicles are increasingly important to our understanding of high-speed aerodynamics.

The X-51 WaveRider UAV is designed to investigate extremely high-speed flight. It is powered by a scramjet (supersonic ramjet), an 'air breathing' jet engine designed to operate at extremely high speeds. In order to generate sufficient power to drive the craft to hypersonic speeds, the engine requires large amounts of oxygen that is obtained by allowing air to pass into the engine. Given the speeds that a scramjet-powered craft

operates at, it could be said to 'ram' air in (hence 'ramjet'), but what is really happening is that the craft is pushing itself into the space occupied by the air.

A narrowing of the intake path causes the airflow to remain at supersonic speeds relative to the engine, effectively compressing the air to increase the amount of oxygen contained in a given volume. This supports the combustion of high-powered JP-7 jet fuel developed originally for the SR-71 reconnaissance aircraft. JP-7 requires so much oxygen to combust that it is possible to use it to put out a cigarette under normal atmospheric conditions.

The scramjet engine can only operate at very high speeds, typically above Mach 4.5 (4.5 times the speed of sound). The WaveRider UAV reaches this speed by first being dropped from a parent aircraft and then boosted using a rocket. This burns out quickly and is discarded, leaving the WaveRider to fly on using its own scramjet propulsion.

Most aircraft obtain lift by way of differential airflow over wing surfaces, but conventional wings would not be appropriate at such high speeds; they would either suffer structural failure or would produce too much drag and prevent the craft reaching high speed. Instead, WaveRider remains aloft by 'riding' the shock waves created by its passage through the air – hence its name. Its small control surfaces are necessary for stability and orientation.

WaveRider achieved Mach 5 on its maiden flight in 2010, and it is hoped that it will eventually be able to fly at Mach 7 or more. This is the forefront of aerodynamics – some spacecraft travel faster during entry to earth's atmosphere but, in terms of sustained powered flight by an air-breathing craft, it is very much an area where experimentation is only just beginning.

The technology required to build hypersonic craft is new – not only must propulsion systems be developed, but

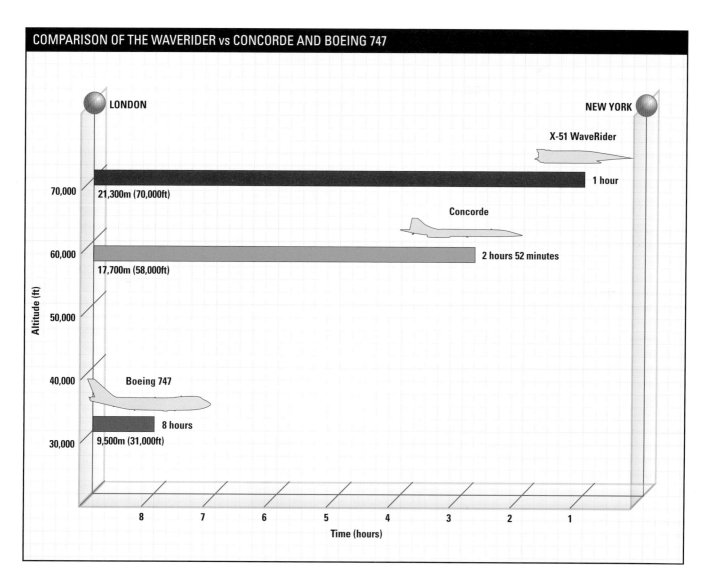

COMPARISON OF THE WAVERIDER vs CONCORDE AND BOEING 747

LONDON

NEW YORK

X-51 WaveRider

1 hour

70,000

21,300m (70,000ft)

Concorde

2 hours 52 minutes

60,000

17,700m (58,000ft)

Altitude (ft)

50,000

40,000

Boeing 747

8 hours

9,500m (31,000ft)

30,000

Time (hours)

8 7 6 5 4 3 2 1

also the metals capable of withstanding the heat generated by hypersonic flight, and constriction techniques for using them. Control technology must also be developed and refined in the light of knowledge gained about the behaviour of air under extreme conditions.

This knowledge can only be gained through hazardous experimentation. WaveRider UAVs have already been lost in tests – any malfunction will likely result in a loss under such extreme conditions. However, knowledge gained has many applications. The military is interested in hypervelocity kinetic-kill weapons, i.e.

missiles that travel so fast that they are all but impossible to intercept and rely on their own mass and velocity rather than a warhead. A Mach 6+ impact delivers an enormous amount of energy to a target and may penetrate quite deeply into earth or concrete, making it effective against bunkers.

At times, there have also been flickers of interest in hypersonic airliners that would dramatically reduce inter-continental flight times. In order to be commercially viable, the technology must be proven to be more or less entirely safe under working conditions. We are

SPECIFICATIONS: X-51 WAVERIDER

Length: 4.3m (14ft); (7.62m (25ft) with booster rocket)
Empty weight: 1,814kg (4,000lb)
Maximum speed: 5794km/h (3600mph) or more
Range: 740km (460miles)
Service ceiling: 21,300m (70,000ft)

very far from that stage as yet, but these early steps with hypersonic demonstrator aircraft may eventually lead to a commercially viable technology. The fact that there is no pilot aboard will surely save lives along the way.

Space drones

Outer space is about the most hazardous environment imaginable. Spacecraft are subject to alternate intense heating from the sun and the cold of vacuum as they pass in and out of the earth's shadow – indeed, opposite sides of a craft may be subject to an enormous temperature differential if one is in direct sunlight. Solar radiation is another severe hazard, along with the possibility of a puncture that can lead to depressurization. Any task that can be carried out remotely exposes astronauts to less risk, especially if it can be done without sending people into space at all.

In the longer term, microgravity is not good for humans; even on a short space mission there are physiological effects that can be serious. Some astronauts become seriously 'space sick' due to the lack of gravity, and it is not possible to tell who will be badly affected without venturing into space to find out. On top of all that, there are the dangers of re-entry and the fact that a space vehicle's propulsion system is capable of exploding and destroying the craft and everyone aboard.

Thus it makes sense to use unmanned vehicles as much as possible. Every manned launch risks human lives; an unmanned craft that suffers an accident is hugely expensive, but at least not such a tragedy. There is also the benefit that humans need air, food, water, heating and cooling systems and space to move

Left: The X-37B is a technology demonstrator, i.e. it was developed to test and develop the technologies that will be used in a new generation of space vehicles. It is designed to be launched by some other vehicle but returns to earth autonomously after its mission is completed.

around in, whereas an unmanned craft can be smaller and lighter or carry more for the same lifting capacity.

Thus while it is immensely exciting to launch heroes into space to perform experiments or crew a space station, it makes sense to use unmanned vehicles as much as possible. Perhaps only a human can carry out certain tasks, but it may not be necessary to have a crew aboard the supply rocket.

X-37B

Most space launches employ one-use rocket systems. As the fuel in each section of the rocket burns out, it is ejected, reducing the weight that has to be lifted by the remaining engines. Only a small part of the overall launch system reaches orbit, and that is usually non-reusable. Thus the entire cost of the lifting system must be paid each time something is placed in orbit.

A reusable space vehicle reduces costs by being available for repeated launches. This is offset by the cost of developing it, but overall a reusable system should enable more missions to be carried out on the same budget. The Space Shuttle was the first successful reusable space vehicle, but since its introduction there have been numerous projects aimed at creating a cost-effective space launch system.

Among these is the X-37 Orbital Test Vehicle, which is designed to be launched vertically in the manner of a conventional rocket, but can make use of aerodynamic flight to return to earth. It is a fully automated 'spaceplane' that began development in 1999, and first flew in 2006. This was not a space mission, but a test intended to investigate the vehicle's flight characteristics and study its automated systems.

The tests were not uneventful. Weather conditions and hardware problems delayed the first free flight, and, when a successful flight was made, the craft was damaged during landing.

Above: X-37B is currently launched by an expendable rocket system, such as Atlas V. Eventually the intent is to create autonomous reusable space vehicles, which can carry payloads to orbit cheaply and safely, and without the requirement to lift people and everything they need to survive in space.

COMPARISON OF THE FIRST SPACEPLANES

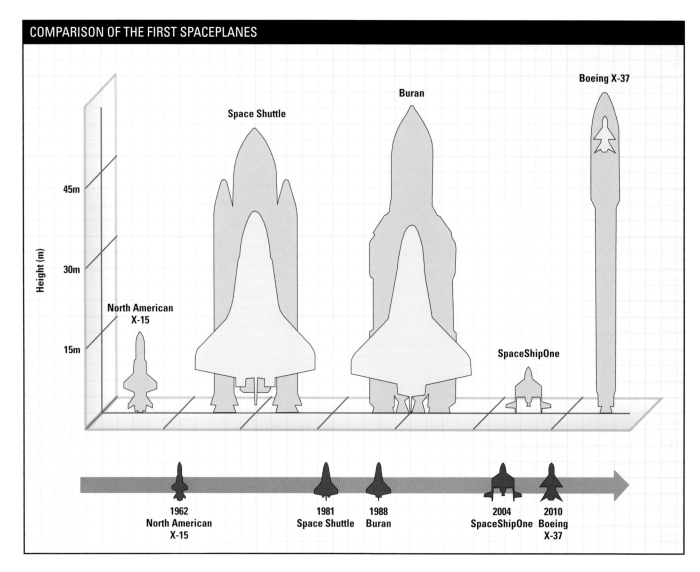

However, the X-37 made its first space mission in 2010 and has been proven capable of making a safe landing. Its missions to date have been of long duration – over 1360 days spread over three missions – which would be an extremely long time for astronauts to be in orbit. Exactly what the Orbital Test Vehicle was doing for so long in space remains classified, but it is likely that it was living up to its name and serving as a testbed for sensors and other equipment. Automated space vehicles are very much a new technology, and clearly there is much to learn.

The original X-37 design was intended to fit in the Space Shuttle's payload bay for launch, although it proved more cost-effective to launch it using a conventional rocket. The X-37A version was used for drop tests, with the X-37B developed from it into a space-capable vehicle.

An enlarged version of X-37, designated X-37C, is under development. This vehicle is capable of carrying a team of astronauts in its mission bay, but will presumably still operate autonomously. This would allow teams of scientists to be carried into orbit without the need for a flight crew. For the same lifted weight a

mission would thus be able to carry more instrumentation, more equipment or more scientific expertise.

It has been speculated that this vehicle could be used for spying, or even as a weapons platform. The possibility exists, of course, but that comment applies to any space launch system. What the X-37 system is more likely to do is to help open the way for commercial operations in space by proving that a low-cost orbital launch system is a real possibility.

Space Station Resupply Vehicles
The cost of the International Space

Above: Automating resupply operations for the International Space Station means that any given resupply mission can carry more, and there is no need to train what might be considered spacegoing delivery van drivers. This, in turn, means that less of the space budget is spent on routine supply runs, and more on the primary mission.

Above: The X-37 vehicle has demonstrated that it can cope with long-duration space missions and the difficulties of piloting a hypersonic craft towards a relatively small landing area. Safety is of paramount importance when dealing with objects that could end up falling from a very great height.

SPECIFICATIONS: X-37B

Length: 8.9m (29ft 3in)
Wingspan: 4.5m (14ft 11in)
Height: 2.9m (9ft 6in)
Loaded weight: 4990kg (11,000lb)
Powerplant: 1 × Aerojet AR2-3 rocket engine (hydrazine), 6600lbf (29.3kN)
Orbital speed: 28,044km/h (17,426mph)
Orbital time: 270 days

Station (ISS) goes beyond the already enormous investment in building it. The station must be resupplied at intervals, and every orbital launch is expensive. Space Shuttle missions were a somewhat cheaper method of delivering supplies to the space station than conventional rockets, but since the curtailment of the shuttle programme this has not been an option.

An automated resupply drone offers one way of reducing costs, since it does

Left: The first mission by the Automated Transfer Vehicle to the International Space Station used a capsule named after the science-fiction writer Jules Verne, and carried priceless original manuscripts of two of his novels. Risking historic documents was a token of confidence in a concept that might have seemed like science fiction not long ago.

away with the need to lift an astronaut to orbit along with all of the air, water and other supplies he will need for the trip and the fuel required to lift his weight. However, the challenges inherent in creating such a craft are not trivial. Even a minor error could result in a collision with the station that, even if it did not destroy it, could result in disaster. Simply nudging the ISS slightly out of alignment could rob it of solar power. or alter its orbit to a degree that might not be easily corrected.

Several automated vehicles have been used to carry supplies to the ISS. The Russian Progress vehicle was specifically developed to support manned space stations and has seen extensive use with the International Space Station. Although not crewed during flight, it is often considered to be a manned spacecraft because the station crew can enter the pressurized forward compartment.

Progress vehicles are no longer in use, and have been replaced by other automated 'space freighters'. These include the European Space Agency's Automated Transfer Vehicle (ATV). This made use of the Progress vehicle's docking system, but carries three times as much cargo. It is similar in that it has a pressurized cargo area that enables the station crew to unload the capsule without the huge encumbrance of a space suit. As a secondary capability, the ATV can provide thrust to the space station to move it into a higher orbit when necessary.

The ATV makes use of GPS guidance, as well as a star-tracking system to navigate to the station, and docks automatically. In an emergency, the crew can order the ATV to abort its approach and initiate an anti-collision protocol. Once safely docked, the capsule is used as an orbital booster from time to time, and serves as a repository for waste. Once full, it is detached and deorbits, burning up in the atmosphere.

The Japanese H-II Transfer Vehicle, as well as the Dragon and Cygnus

Above: The H-II Transfer Vehicle, or Kounotori, is the Japanese equivalent to the European Automated Transfer Vehicle. The concept of automated space trucks is sufficiently well proven that various models are beginning to appear, perhaps marking the beginning of a new era in unmanned space operations.

Transfer Vehicles that are built by private contractors rather than national agencies, use slightly different docking procedures. The craft proceeds automatically and autonomously through a series of waypoints, but must be given permission to proceed to the next by the crew of the space station.

These 'space freighters' have been in use for some years, and, although it may be a while before automatic space vehicles become routine and commonplace, they have proven that the concept is viable. As commercial exploitation of space expands, it is likely that increasing numbers of 'space drones' will enter service. Presumably this will require a new acronym: USV (Unmanned Space Vehicle) or UOV (Unmanned Orbital Vehicle) seem likely. The term, UAV, on the other hand, implies operations within the atmosphere and may not be appropriate for drones that operate outside it.

The future

Drone operations were not really possible on a widespread scale until recently, but, now that the technology is available at an affordable cost, the only remaining question is whether or not drones will prove useful enough to create a viable industry. In order for this to happen, enough users in enough different sectors must begin to routinely use drones of one sort of another so that large-scale construction becomes and remains a profitable business. Is this likely?

Above: Although the demise of the manned fighter in favour of drones and missiles has been predicted since the 1970s, for the foreseeable future a piloted aircraft will remain capable of outperforming a drone. This ensures the survival of advanced manned fighters, such as the F-22 Raptor, and probably the next generation of combat aircraft, too.

The answer to that question is a resounding 'Yes'. Drones are available to meet a wide range of requirements from military to scientific, and for recreation, too. From relatively humble beginnings we have seen drones gain ever more impressive capabilities that could allow them to displace manned craft from some applications.

It was suggested four decades ago or more that manned combat aircraft had no future; that the guided missile and the unmanned aircraft would render pilots obsolete. The technology to do this did not exist at the time, but it does now. However, it still remains questionable whether autonomous combat aircraft are a good idea, either operationally or ethically. No hardware or software was ever without a bug or two, making the wisdom of sending out armed robotic aircraft dubious. Besides, it remains to be seen whether a machine can really outperform a human in all aspects of combat operations.

Commercial Air Traffic
The situation may be somewhat different in the realms of commercial air traffic.

Change of any sort is usually regarded with suspicion, and it may be some time before people are willing to trust their safety to a fully automated aircraft – even if it is shown to be a more capable pilot than any human. It was found that an image of a human driver made people feel better about using automated public transport in certain cities, which perhaps says a lot about human psychology.

That said, once partial acceptance occurs, many initially strange concepts can become commonplace. It is likely that widespread acceptance of automated passenger flights will come – if it comes at all – in a series of small steps starting perhaps with robotic package delivery systems, and progressing to commercial cargo transportation or perhaps automated resupply of remote areas. This is an area where drones could be used to cut costs, and perhaps risks, by using automated deliveries to oil platforms or scientific stations in Arctic or Antarctic regions.

Hostile Environments

Automated vehicles are also a possibility for exploration of extremely remote and hostile environments. Rather basic versions have been used to conduct experiments and limited exploration on the moon and even Mars. More advanced drones might be able to deal with hazards and setbacks somewhat better, although for such long-duration missions the main problems are related to providing power rather than complexity of circuitry.

It is possible that in the fairly near future we may see a 'Mars Flyer'

Right: The landing of vehicles, such as the Curiosity Rover, on Mars represents the current pinnacle of drone technology. Sending a vehicle all the way to Mars was an incredible achievement, and landing it within 2.5km (1.5 miles) of the aim point is almost unbelievable. It has since sent back a wealth of data about its new environment.

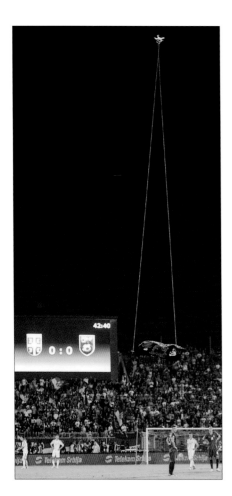

Left: In October 2014, a drone was used to fly the Albanian flag over the pitch during a Serbia versus Albania clash at the Partizan Stadium in Belgrade. The match was abandoned after home defender Stefan Mitrovic pulled down the flag carried by the drone – leading to chaos on the field and in the stands.

drone that is either dropped into the atmosphere of Mars from orbit or launched from a platform landed on the ground. NASA's experimentation with high-altitude drones has shown that it is possible to fly in very thin gas, such as that found in the atmosphere of Mars. Given a lower gravity and similar atmospheric density, a flying exploration platform seems plausible.

Remote exploration of the seabed or hazardous environments, such as thermal vents, deep underwater caves, volcanoes and under pack ice, are all possible. A remote-controlled vehicle might not be able to receive control signals in some areas, which would also preclude sending data back. However, a 'buddy' system could be used, with a data drone following the exploration

unit and receiving streamed data from it. If the exploration drone signal stops, e.g. because it has been crippled or destroyed by some hazard, then the data unit turns back and brings its findings to the operators.

It may be that the final limit on drone technology is what we are willing to permit. Already ethical questions have been raised about armed drones and whether it is acceptable to give control over aircraft entirely to machines. Advancing technology is forcing us to confront other issues as well.

Safety Concerns

The availability of low-cost drones as recreational items raises questions about safe operation. Some people show a quite incredible lack of common sense, and without safe-operations training (or perhaps despite it) can be expected to fly drones in completely inappropriate places. High-speed flying objects pose something of a hazard to people nearby in the event of control loss or irresponsible flight, and it is possible that motorists or those working in hazardous positions (such as at a height) could be distracted or even deliberately harassed with drones.

There is also a possibility of drones being flown in front of moving trains or into restricted airspace. The latter could pose a serious hazard to aircraft and could disrupt the operation of an airport. Legal penalties for misuse of drones might be a partial deterrent, but some operators will simply not think – and others might see a drone as a means to cause harassment as a way of making a

point. Whether relatively benign protest about something the individual dislikes or as part of an act of terrorism, the possibility of misuse is a real one.

Surveillance Concerns

Many drones have cameras on them, which also raises privacy and security questions. Careers can be ruined by one photograph, and while those engaged in illegal activity perhaps deserve little sympathy, it is not unreasonable to expect privacy when at home or in a secluded holiday location. Camera-equipped drones offer unscrupulous photographers the chance to obtain 'candid' images of people who incorrectly thought they were in private. Given the market for such pictures, misuse is all but inevitable.

Drones could also be used for harassment and spying in a domestic or neighbourly dispute, or to obtain information for use in committing a crime. Since there is no need to license drone use and few limitations on buying them, there is little deterrent to such use and few ways to legally prevent it.

Drone misuse will likely result in court cases in the next few years dealing with issues such as who owns the airspace above your garden? This is not really a concern when it deals with airliners or the occasional police helicopter passing over, but what of a drone hovering at the height of your upstairs windows? Is it legal to take a swipe at a drone hovering in front of your face and being deliberately used to annoy you?

The increasing availability of drones, and their ever more widespread use,

Right: Whilst responsible use is almost certainly harmless, a reckless or thoughtless drone operator could end up causing a serious hazard. A drone flown into traffic or in front of a train may cause a lot of damage, whilst the whirling blades of many multirotor types could cause injury if an accidental collision occurred.

will inevitably raise legal questions. Some of these are predictable and could be considered ahead of time, at least in a general way. Others will be totally unexpected. This is the nature of innovative technology and new capabilities. The only thing we can say for certain is that drones have proven their usefulness and that the necessary technology is in place. In other fields, we have seen that once that happens, growth and progress occur fast and often in unexpected directions.

Right: Drone misuse could become a real problem. A UAV hovering out in the street or in some other area of 'public airspace' could point its camera into a window or over hedges and obtain inappropriate images. Creating legislation to prevent this is likely to be complex, if it is even possible.

Glossary

ATV Automated Transfer Vehicle

AUV Autonomous Underwater Vehicle

Avionics A term that encompasses the electronics found within an aircraft, usually in the cockpit. For UAVs, avionics are housed in a nose blister

CALCM Conventional Air-Launched Cruise Missile

COMINT Communications Intelligence

COTS Commercial-Off-The-Shelf

EOD Explosive Ordnance Disposal

Fixed-wing Conventional aircraft that have fixed wings to generate lift as opposed to rotary wing aircraft that generate lift from one or more rotors powering wings that turn in the horizontal plane

FLIR Forward-Looking Infrared. Heat-sensing equipment that scans the path ahead to detect heat from objects such as vehicle engines

FPV First Person View

Gimbal An attachment for a drone that keeps another mounted device (most often a camera) level and stable

GPS Global Positioning System. A system of navigational satellites

Gyroscope A device that keeps a UAV's flight balanced while in midair

HALE High-Altitude, Long-Endurance

HAPS programme High Altitude Pseudo-Satellite programme

HARM Homing Anti-Radiation Missile

HMMMV High Mobility Multipurpose Military Vehicle

IDF Israeli Defence Force

IED Improvised Explosive Device

ISTAR Intelligence, Surveillance, Target Acquisition, Reconnaissance) platform

Laser rangefinder A rangefinder, which uses a laser beam to determine the distance to an object

LIDAR Laser Imaging Radar

LPI Low-probability-of-intercept radar equipment

MALE Medium Altitude Long-Endurance

Maximum Takeoff Weight The maximum weight due to design or operational limitations, at which an aircraft is permitted to take off

MOUT Military Operations in Urban Terrain

MP-RTIP Multi-platform radar technology insertion programme

MUM-T Manned-Unmanned Teaming

No Fly Zone Areas where flying a drone is restricted by government regulations. A place where a drone could interfere with an aircraft or record sensitive information make up most of these no fly zones

PARM Persistent Anti-Radiation Missile

Payload The weight of passengers and/or cargo

QUADCOPTER A type of aircraft propelled by four rotors

RAPTOR Responsive Aircraft Program for Theater Operations

RATO (Rocket Assisted takeoff) booster

Reconnaissance Exploring beyond the area occupied by friendly forces to gain vital information about natural features and enemy presence in a given area for later analysis and/or dissemination

ROV Remotely Operated Vehicle

SAM Surface-to-air missile

SAR Synthetic Aperture Radar

SATCOM Abbreviation for Satellite Communications

SCUD Missiles A series of tactical ballistic missiles developed by the Soviet Union during the Cold War

SEAD Suppression of Enemy Air Defences

Service Ceiling The maximum usable altitude of an aircraft

SIGINT Signals Intelligence

SOCOM Special Operations Command

SONAR A technique that uses sound propagation (usually underwater, as in submarine navigation) to navigate, communicate with or detect objects on or under the surface of the water, such as other vessels. An Acronym for SOund Navigation And Ranging

Stealth Technology Technology applied to aircraft or fighting vehicles to reduce their radar signatures

Tank Plinking The practice of using precision-guided munitions to destroy artillery, armored personnel carriers, tanks, and other targets

TERCOM Terrain Contour Matching

Thermal Imaging Equipment fitted to an aircraft or fighting vehicle which typically comprises a telescope to collect and focus infra-red energy emitted by objects on a battlefield, a mechanism to scan the scene across an array of heat-sensitive detectors, and a processor to turn the signals from these detectors into a 'thermal image' displayed on a TV screen

UAV Unmanned Aerial Vehicle

UCLASS programme Unmanned Carrier-Launched Airborne Surveillance and Strike programme

Ultrasonic Sensor A device for detecting objects using radio and sound waves

VHF Acronym for Very High Frequency

VTOL Acronym for Vertical Takeoff and Landing

Waypoint A reference point in physical space that aids aircraft navigation

Index

Page references in *italics* refer to figure: page numbers in **bold** refer to material in boxes

Picture Credits